SpringerBriefs in Molecular Science

Biobased Polymers

Series editor

Patrick Navard, Sophia Antipolis cedex, France

Published under the auspices of EPNOE*Springerbriefs in Biobased polymers covers all aspects of biobased polymer science, from the basis of this field starting from the living species in which they are synthetized (such as genetics, agronomy, plant biology) to the many applications they are used in (such as food, feed, engineering, construction, health, ...) through to isolation and characterization, biosynthesis, biodegradation, chemical modifications, physical, chemical, mechanical and structural characterizations or biomimetic applications. All biobased polymers in all application sectors are welcome, either those produced in living species (like polysaccharides, proteins, lignin, ...) or those that are rebuilt by chemists as in the case of many bioplastics.

Under the editorship of Patrick Navard and a panel of experts, the series will include contributions from many of the world's most authoritative biobased polymer scientists and professionals. Readers will gain an understanding of how given biobased polymers are made and what they can be used for. They will also be able to widen their knowledge and find new opportunities due to the multidisciplinary contributions.

This series is aimed at advanced undergraduates, academic and industrial researchers and professionals studying or using biobased polymers. Each brief will bear a general introduction enabling any reader to understand its topic.

*EPNOE The European Polysaccharide Network of Excellence (www.epnoe.eu) is a research and education network connecting academic, research institutions and companies focusing on polysaccharides and polysaccharide-related research and business.

More information about this series at http://www.springer.com/series/15056

Nurudeen A. Oladoja · Emmanuel I. Unuabonah
Omotayo S. Amuda · Olatunji M. Kolawole

Polysaccharides as Green and Sustainable Resources for Water and Wastewater Treatment

 Springer

Nurudeen A. Oladoja
Hydrochemistry Research Laboratory,
    Department of Chemical Sciences
Adekunle Ajasin University
Akungba-Akoko, Ondo
Nigeria

Emmanuel I. Unuabonah
Environmental and Chemical Processes
    Research Laboratory, Department of
    Chemical Sciences
Redeemer's University
Ede, Osun
Nigeria

Omotayo S. Amuda
Environmental Technology Research
    Laboratory, Department of Pure
    and Applied Chemistry
Ladoke Akintola Univeristy
Ogbomosho, Oyo
Nigeria

Olatunji M. Kolawole
Infectious Diseases and Environmental
    Health Research Laboratory, Department
    of Microbiology
University of Ilorin
Ilorin, Kwara
Nigeria

ISSN 2191-5407   ISSN 2191-5415   (electronic)
SpringerBriefs in Molecular Science
ISSN 2510-3407   ISSN 2510-3415   (electronic)
Biobased Polymers
ISBN 978-3-319-56598-9   ISBN 978-3-319-56599-6   (eBook)
DOI 10.1007/978-3-319-56599-6

Library of Congress Control Number: 2017939088

Printed on acid-free paper

This Springer imprint is published by Springer Nature
The registered company is Springer International Publishing AG
The registered company address is: Gewerbestrasse 11, 6330 Cham, Switzerland

Foreword

Polysaccharides are collections of different carbohydrates that occur naturally. The skeletal framework is characterized by the composition of either a type of repeat unit of monosaccharide (i.e. homopolysaccharides or homoglycans; e.g. starch, cellulose) or a combination of different repeat units (i.e. heteropolysaccharides or heteroglycans; e.g. agar, alginate, carrageenan). Polysaccharides are among the most abundant and prevalent renewable natural resource that is found in every part of the globe. This fascinating and almost inexhaustible biopolymer possesses remarkable chemical and physical attributes which broaden its application.

The industrial relevance of polysaccharides date back to thousands of years and has currently gained interesting and valuable applications in the biomedical, engineering, food and environmental fields. Polysaccharides can be obtained from some sources which include seaweeds, plants, bacteria, fungi, insects, crustacean, animals and even humans. As a result of the rich chemistries of polysaccharides, the structural framework can be tuned, and tailor-made, through genetic and structural engineering, to serve desired purposes.

An overview of the recent trends in environmental engineering showed that the shift towards the development of green, sustainable and eco-friendly water and wastewater treatment technologies had made the use of natural polymers more preferable to synthetic polymers. This has been attributed to their inert, safe, non-toxic, biocompatible, biodegradable, low cost, eco-friendly and abundant nature. Consequent upon the inherent auspicious features of polysaccharides, as a green bioresource, the use of polysaccharides in some tertiary unit processes in water and wastewater treatment operations have been promoted, and studies in this field have continued to burgeon. In the present treatise, an in-depth review on the use of polysaccharides, as a green and sustainable resource, for water and wastewater treatment is presented. Premised on the rich laboratory and field experiences of the authors, whose research focus is on the development of sustainable substitutes for conventional materials used for water and wastewater treatment, a systematic review of this abundant resource (polysaccharides), as a sustainable technological material of the present and the future, is presented herein.

The current critical review hinges on the use of polysaccharides as operational materials in two tertiary water treatment operations (i.e. coagulation/flocculation and adsorption-based water treatment methods). The operational principles, material requirements, mechanistic insights into the underlying mechanism(s) of operations and the possible structural modifications that have been performed on polysaccharides to optimize and tailor their performances to suit specific applications were systematically and critically discussed. In addition to the provision of the laboratory and field experiences of the authors, relevant literature reviews were also conducted to juxtapose experiences from different divides.

Contents

About the Authors

Dr. Nurudeen Abiola Oladoja teaches chemistry and conducts research in the field of Hydrochemistry in the Department of Chemical Sciences, Adekunle Ajasin University, Akungba Akoko, Nigeria. He is currently the Head of the Hydrochemistry Research Laboratory in Adekunle Ajasin University, Akungba Akoko, Ondo State, Nigeria. He holds a Ph.D. (2003) in Environmental Chemistry (Hydrochemistry), a Masters in Industrial Chemistry (1994) from the University of Benin, Nigeria and a Bachelor of Science in Chemistry Education (1991) from the University of Lagos, Nigeria. He was a Postdoctoral Fellow (CAS/TWAS Postdoctoral Fellowship 2008/2009) at the Research Centre for Eco-Environmental Sciences, Chinese Academy of Sciences, Beijing, China; Visiting Research scientist (USM/TWAS Visiting Research Fellowship, 2010/2011), School of Chemical Engineering, Universiti Sains Malaysia, Malaysia; Visiting Professor (Alexander von Humboldt Fellowship for Experienced Researchers 2013–2015), Chair of Urban Water Systems Engineering, Technical University of Munich, Germany. His research focus in the field of hydrochemistry include: Development of substitutes for the conventional materials in adsorption and coagulation-flocculation based water treatment technologies; Development of green and sustainable appropriate technologies for water and wastewater treatment; Synthesis and characterization of materials for advanced oxidation and reduction process for pollutant attenuation in water and wastewater; Resource recovery from wastewater; Development of strategies for eutrophication control and management. His research work has been financially supported, through research grants, by International Foundation of Science, Sweden (2008 and 2012); Adekunle Ajasin University (2010 and 2013); Alexander von Humboldt Foundation, Germany (2013–2015, 2015 and 2016), Tertiary Education Trust Fund, Nigeria (2016) and The Commonwealth (Commonwealth Travel Grant, 2016/17). He has published over seventy (70) articles in reputable international Journal. He wrote Chap. 2 and co-Authored Chap. 1 in this book.

Dr. Emmanuel I. Unuabonah is an Associate Professor in the Department of Chemical Sciences, College of Natural Sciences, Redeemer's University, Nigeria and also the Head of the Environmental and Chemical Processes Research Group in the University. His research is focused on the preparation and characterization of functional composite materials from local materials and agrowastes for sustainable treatment of water. He is also developing new micro-mesoporous, bacteriostatic and bacteriocidal composite materials for adsorptive and photocatalytic purposes. He has published several articles in international peer-reviewed journals and has won several international awards and research fellowships including the TWAS-ROSSA Young Scientist Award, SCOPE–Zhongyu Young Scientist Environmental Award (Environmental Technological Innovations category), African Union-TWAS Young Scientist Award in Basic Sciences, Technology and Innovation, the prestigious Alexander von Humboldt (AvH) Postdoctoral Research Fellowship at the Institute of Chemistry, Universitaet Potsdam, Potsdam, Germany and the Redeemer's University Researcher of the Year (2013). A work from his research group won the *Dhirubhai Ambani Chemical Engineering Innovation for Resource-Poor People* Award category for his University at the Institution of Chemical Engineers (IChemE) Global Awards (2014), United Kingdom. His research group also won The World Academy of Sciences (TWAS) Research Group grant (2014). He is the Founding President and a Fellow of the Nigerian Young Academy, an alumnus of the TWAS Young Affiliate program and the Global Young Academy (GYA) and a Fellow of the African Science Leadership Program (ASLP). He has supervised and co-supervised several M.Sc. and Ph.D. students. He wrote Chap. 4 and co-authored Chap. 1 in this book.

Prof. Omotayo Sarafadeen Amuda an astute researcher and a prolific scientist has been teaching and conducting research for 18 years in the University system. He is a Professor of Chemistry at Ladoke Akintola University of Technology, Ogbomoso, Nigeria. He is the Head of the Envirommental Technology Research Laboratory, Ladoke Akintola University, Ogbomoso, Oyo State, Nigeria. He has contributed to chapters in books published by Taylor and Francis UK, Springer, USA and World Scientific Publishing Company Limited, Singapore. He has about seventy academic publications/conference proceedings in reputable national and international publishing outlets.

He won the first position at the first Ladoke Akintola University of Technology Research and Development Fair, organized by the Academic Planning Unit of the University, November, 2006. He is a Fellow of the International Chartered World Learned Society, USA and he is a recipient of "Salute to Greatness" Awards from this same society. He is a Knight of Chartered World Order of the Knights of Justice of Peace Inc. USA. He was a visiting Lecturer and acting Head, Department of Chemical Sciences at Wesley University of Science and Technology, Ondo, Nigeria. He is currently a visiting Professor at the Department of Chemical Sciences, Osun State University, Osogbo, Nigeria.

His current research interest is in the synthesis and characterization of TiO_2/Ag doped activated carbons for photocatalytic and antibacterial applications in wastewater treatment. He wrote Chap. 3 in this book.

Dr. Olatunji Matthew Kolawole is Associate Professor of Medical and Environmental Microbiology (Virology) at the University of Ilorin, Nigeria, and is currently the group head of Infectious Diseases and Environmental Health Research Group, at the same university. He is an astute and prolific researcher with over 80 published articles in scientific peer-reviewed national and international journals. He conducts researches in the areas of infectious diseases and toxicology, water quality and surveillance in relation to humans and animal diseases in the tropics. He is currently the National coordinator, Poliovirus Laboratory Containment Programme and a member of the National Taskforce on Poliovirus Laboratory Containment in Nigeria. He is a recipient of several international fellowship awards and research grants. He has supervised and graduated undergraduates, M. Sc. and Ph.D. students in the Department of Microbiology, University of Ilorin, Nigeria. He is an external examiner to some notable universities in Nigeria. He wrote Chap. 5 of this book.

Chapter 1
Operational Principles and Material Requirements for Coagulation/ Flocculation and Adsorption-based Water Treatment Operations

1.1 Introduction

Water and wastewater treatment operations require different unit processes or operations. These unit processes (e.g. screening, chemical pre-treatment, coagulation/ flocculation, filtration, disinfection, adsorption using granular activated carbon, aeration, reverse osmosis, electrodialysis, etc) are combined to form a single treatment system. The treatment system may contain all of the unit processes or a combination of any of them. Thus, one or more of these operations may be used to treat water or wastewater. Based on the precise nature of the treatment and the extent of treatment that can be guaranteed by a particular treatment procedure, water and wastewater treatment procedures are broadly classified into primary, secondary and tertiary methods of treatment. Primary treatment (e.g. screening, pre-sedimentation, and chemical addition) provides a broad degree of treatment and the specificity is low. Secondary treatment (e.g. coagulation-flocculation, filtration, disinfection) provides a higher magnitude of treatment and is more specific than the primary treatment while the tertiary treatment methods (e.g. adsorption, reverse osmosis, etc.) are often applied when a higher degree of treatment is required, and it is often pollutant specific.

On a broader view, adsorption is an enrichment of chemical species from a fluid phase on the surface of a liquid or a solid [1]. In water and wastewater treatment operations, adsorption has proven to be an efficient and reliable tertiary treatment procedure for the removal of pollutants from aqueous system. Coagulation-flocculation (CF) is a chemical clarification process that agglomerate colloidal and suspended matters. It often serves as either a primary or secondary mode of treatment while adsorption is strictly a tertiary treatment procedure.

© The Author(s) 2017
N.A. Oladoja et al., *Polysaccharides as a Green and Sustainable Resource for Water and Wastewater Treatment*, Biobased Polymers,
DOI 10.1007/978-3-319-56599-6_1

1.2 Operational Principles of Adsorption and Coagulation/Flocculation

1.2.1 Adsorption Based Water Treatment Operations

Adsorption-based water treatment operation is the process through which a substance or a solute, initially present in the aqueous phase, is removed from that phase by accumulation at the interface between the aqueous phase and a solid phase. The ability of the solid phase to perform this function is hinged on the fact that such solid surfaces are characterized by active, energy-rich sites which enable them to interact with solutes in the aqueous phase because of their specific electronic and spatial properties [1]. Typically, active sites have different energies (i.e. the surface is energetically heterogeneous). In the course of the adsorption process, if the process variables (e.g. concentration, temperature, pH, ionic strength) are altered, the reverse of the adsorption process could be triggered and the adsorbed species get released from the surface and transferred back into the aqueous phase. This reverse process of adsorption is known as desorption. The solid phase that provides the surface as the platform for adsorption is termed the *adsorbent* while species that are adsorbed on the adsorbent are referred to as *adsorbates*. A brief description on conventional adsorbents of industrial significance are presented below:

(i)   Activated carbon: This is a char-like carbonaceous material with high surface area. Powdered and granular activated carbons are the two types of activated carbon which are used differently in adsorption procedures because of their different characteristics.
(ii) Silica gel: This is a hard, granular, porous material that is made via precipitation of silica from sodium silicate solution in acid.
(iii) Activated alumina: This is produced through the activation of aluminium oxide at high temperature and is used primarily for moisture adsorption.
(iv) Aluminosilicates (also known as molecular sieves): They are porous synthetic zeolites or materials that are used mainly in separation processes.

Adsorption is a century old drinking water treatment procedure that uses activated carbon as the adsorbent. Primarily, the process is aimed at the removal of taste and odour imparting compounds from the aqua stream. Recently, the conventional adsorbent (i.e. activated carbon) has proven to be efficient in the removal of an array of organic micropollutants (e.g. phenols, chlorinated hydrocarbons, pesticides, pharmaceuticals, personal care products, corrosion inhibitors, etc.) from aqueous matrices [1]. Adsorption is a tertiary wastewater treatment procedure for polishing partially treated water or wastewater before being stored or distributed for consumption in the case drinking water, and for final discharge in the case of wastewater. Soluble organic substances that are recalcitrant to biodegradation by the conventional biological treatment or which are not easily precipitated by the chemical clarification process are often removed by adsorption onto activated carbon. In conventional adsorption-based water treatment operation, activated carbon is applied either as powdered activated carbon (PAC) in slurry reactors or as granular activated carbon (GAC) in fixed-bed reactors. The particle size of PAC is smaller than 74 µm (the typical range is 10–50 µm), and the material is commonly used by direct addition. For GAC, the particle size is larger than 100 µm (the typical range is 0.4–2.5 µm), and the material is commonly used in fixed bed reactors. Consequent upon the fact that adsorption is a surface phenomenon, the adsorption capacity of activated carbon often correlates with a very high surface area per unit volume.

Based on the nature of interaction that occurs between adsorbents and adsorbates, adsorption is broadly classified as physical adsorption (physisorption) or chemical adsorption (chemisorption). Physical adsorption occurs when the solute is held onto the adsorbent by van der Waals attractive force, while chemisorption occurs when the solute is held onto the adsorbent via specific chemical bonds. The specific chemical bond that occurs in chemisorption can be covalent or ionic in nature. The occurrence of the two types of adsorption (i.e. physisorption and chemisorption) is sometimes experienced simultaneously. Both types of adsorption have different features that differentiate them, and these features are often sought out for to pinpoint the actual type of adsorption that took place in an adsorption operation.

The relationship between the equilibrium concentration of a solute in the aqueous phase and the solid adsorbent phase, at any constant temperature, is called an *adsorption isotherm*. The equilibrium adsorption isotherm is fundamentally important in the design of an adsorption-based water treatment system. The results obtained from the equilibrium isotherm analysis suggests the adsorption capacity of the adsorbent while the different adsorption isotherm equation are characterized by certain constants whose values express the surface properties and affinity of the adsorbent in use. The equilibrium adsorption isotherm is conventionally expressed in the form of two generally known mathematical equations derived by Langmuir (Langmuir equilibrium isotherm equation) and Freundlich (Freundlich equilibrium isotherm equation). These equations are conventionally used to analyse the data obtained from a typical adsorption operation. However, recently, other equilibrium isotherm equations have been developed using these two conventional equations as foundations. The results obtained from the equilibrium isotherm analysis provide an inkling into the underlying mode of the adsorption process.

In the design of adsorption-based water treatment system, the kinetic analysis of the adsorption process is usually analysed to gain valuable insights into the reaction pathways and mechanisms of adsorption. The kinetic analysis helps to define how adsorption rates depend on the concentrations of solute in a solution and how rates are affected by adsorption capacity or by the character of the adsorbent [2]. Thus, the prediction of the rate at which the solute is removed from aqueous solutions established the residence time required for the completion of the process of adsorption. The different kinetic models that have described the reaction order of adsorption systems can be classified into two viz:

(i) Adsorption rate equation based on the adsorbent's capacity for a solid-liquid system.
(ii) The one whose adsorption rate is based on solution concentration.

The kinetic equation whose adsorption rate is based on solution concentration has been identified [3] to include first order [4] and second-order [5] reversible ones, and first-order [6] and second-order [7] irreversible ones, pseudo-first-order [8] and pseudo-second-order ones [1]. The kinetic equations whose reaction orders are based on the capacity of the adsorbent have been identified [3] to include Lagergren's first-order equation [9], Zeldowitsch's model [10], and Ho's second-order expression [11–14].

1.2.2 Coagulation/Flocculation Treatment Operations

Coagulation/flocculation (CF) is a chemical clarification procedure that involves the use of chemicals to remove, by precipitation, impurities from water matrix. Specifically, CF work to destabilize particles and agglomerate dissolved particulate matter. Coagulation results from the addition of a coagulant to the aqueous phase to be treated. Commonly used coagulants are Alum (aluminium sulphate), Sodium Aluminate, Ferric sulphate, Ferrous sulphate, Ferric chloride, and organic polymers. CF operations remove inorganic and organic substances, taste and odour impacting substances, phosphates and biological species. CF produces water that is aesthetically acceptable, properly disinfected and amenable to filtration through sand filters.

Before the advent of synthetic organic polymer coagulants, the metal-based coagulants (e.g. alum, iron (II) and iron (III) salts, slaked lime, quick lime or the mixtures of the chemicals) were the classical and conventional coagulants used in water treatment. An overview of the underlying operational mechanisms of the metal-based coagulants showed that in the aqua matrix, the hydrolysis of the metal salt occurs rapidly to form various cationic species, which are adsorbed by negatively charged particles and caused charge neutralization. At low coagulant dosage, the operational mechanism of the CF process has been ascribed to the charge neutralization mechanism. At sufficiently high coagulant dosage, a process occurs which is designated as "sweep flocculation" and has been attributed to be the underlying mode of the CF process. Sweep coagulation occurs through the precipitation of metal hydroxide. During the formation of the metal hydroxides, the turbidity causing agent (i.e. colloidal particles arising from clay and silt, microorganisms and vegetable material) are trapped in these precipitates.

Another class of coagulants is the polymer-based coagulants whose operational mechanism defers from that of metal-based coagulants. The CF mechanism for this class of coagulants occur via destabilization by bridging, when segments of a polymer chain get adsorbed on more than one particle, thereby linking the particles together. It has been reported that the flocs produced via bridging flocculation are stronger than those formed when particles are destabilized by simple salts [15, 16].

Thus, aggregates formed by polymeric flocculants are significantly more resistant to breakage [17]. An overview of the role of polyelectrolytes in CF operation showed that they enhance floc settling rate, improve process economy, enhances the quality of the water and produce sludge with improved quality characteristics [15–17].

1.3   Material Requirements for Adsorption and Coagulation/Flocculation

1.3.1   Adsorption Based Treatment Operation

Consequent upon the fact that adsorption is a surface phenomenon, the fundamental consideration in the choice of material to serve as an adsorbent is the surface chemistry of the adsorbent. The surface chemistry of an adsorbent is an important factor that determines the possible mode of interaction between the adsorbent and the adsorbate to be removed from the water matrix. Adsorbents take a broad range of chemical forms and different geometrical surface structures, and this is reflected in the range of their applications in industries and laboratory practices [18]. Irrespective of the source of an adsorbent (i.e. either natural or synthetic), adsorbents can be broadly classified, by their chemical composition, as organic or inorganic based materials and these two classes of adsorbents cut across the different sources. Conventionally, activated carbon is the adsorbent of choice in the removal of dissolved organics and inorganics (as pollutants) in water treatment operations, but the inherent shortcomings have engendered the search for suitable substitutes for this conventional adsorbent. The materials that have been studied as substitutes for this conventional adsorbent include clay minerals, carbonaceous adsorbents, polymeric adsorbents, natural and synthetic zeolites, biopolymers, metal oxides, wastes and by-products. Several selected polysaccharide-based adsorbents that has been investigated as substitutes to the conventional adsorbents discussed in Chap. 4.

A comparative analysis of the performance and operational efficiency of adsorbents derived from different sources showed that synthetic adsorbents (also known as engineered adsorbents) exhibit better adsorption capacities and are consistent in their operational efficiency and stability. They are produced under strict quality control and show nearly constant properties. Often, their adsorption profile towards a broad variety of adsorbates is predictable, and recommendations for their applications can be derived from scientific studies and producers information [19]. Albeit, engineered adsorbents could be sometimes expensive. Comparatively, the adsorption capacities of natural (i.e. biogenic or geogenic) adsorbents are much lower, and the properties are subject to stronger variations [19]. They might be interesting to consider due to their low cost and ubiquity, but in most cases, studies on these materials are limited to very specific applications and not enough information are available for a generalized experience and a final assessment [19].

The historical perspective of the evolutional trends of adsorption science and technology linked the development of the adsorption technique to the various types of adsorbents that have been tested and used [18] thus: before World War I the focus was on carbon adsorbents; during the period between World War I and World War II the focus was on active carbons; silica acid gels and aluminium oxides, but after World War II revolutionary progress was made owing to discovery and application of synthetic zeolites [20, 21].

Consequent upon the need to guarantee the safety of the final product in drinking water treatment industries, the adsorbent used must be of very high-quality standards and must be certified by the relevant standard organization. On this note, the number of possible adsorbents is limited and comprises basically of commercial activated carbons and oxidic adsorbents.

1.3.2 Coagulation/Flocculation Operations

Metal salts and synthetic organic polymers are the primary coagulants that have been used in CF operations. Taken all forms of coagulants into consideration, the underlying mechanism of CF that has been identified and proven by researchers are

   (i)   Double-layer compression
  (ii)   Adsorption and bridging
 (iii)   Charge neutralization
 (iv)   Sweep coagulation.

Considering the structural framework and chemical features of the different coagulants, double-layer compression, charge neutralization and sweep coagulation mechanisms are synonymous with inorganic coagulants while charge neutralization and adsorption and bridging mechanisms are synonymous with polymeric coagulants.

For a material to function as a coagulant, it must possess features that would enable it to operate using any of the identified coagulation mechanisms. Some of these features are:

   (i) **High charge density**—This is required to function in double layer compression or charge neutralization mode of coagulation mechanism.
  (ii) **Macromolecular skeletal framework**—This is needed for the adsorption and bridging mechanism to take effect.
 (iii) **Formation of insoluble species**—This is required for sweep coagulation process to take effect.

The materials that have been used as coagulant include metal salts (e.g. Alum (aluminium sulphate), sodium aluminate, ferric sulphate, ferrous sulphate, ferric chloride) and polymeric substances (i.e. polyelectrolytes). In water industries, organic polymers have been utilized for at least four decades [22]. They are mostly water-soluble linear polymers of very high molecular weights (MW) [15]. The group of these organic polymers that are charged, when hydrolyzed, are referred to as "polyelectrolytes", and they possess many characteristic features of their own. The list of selected substitutes to the conventional coagulant that has been studied are presented in Chap. 2.

1.4 Polysaccharides—A Brief Overview

Polysaccharides are polymeric carbohydrate molecules composed of long chains of monosaccharide units bound together by glycosidic linkages and on hydrolysis give the constituent monosaccharides or oligosaccharides. They range in skeletal structure from linear to highly branched. Examples include storage polysaccharides such as starch and glycogen and structural polysaccharides such as cellulose and chitin.

Starch is one of the most abundant polysaccharides, and it is present in plants as energy storage material. It is made up of mixtures of two polyglucans, amylopectin and amylose, but they contain only a single type of carbohydrate, glucose. Chitin is a naturally abundant mucopolysaccharide extracted from crustacean shells, which are wastes products of seafood processing industries. Chitin is the second most abundant polysaccharide in nature, after cellulose, but it is the most abundant amino polysaccharide.

Some of the features that make polysaccharides to be unique as raw materials are

(i)   Abundance as natural polymers (i.e. biopolymers)
(ii) They are cheap materials (low-copolymers)
(iii) The pervasiveness
(iv) It is a renewable resource
(v)   It is a stable and hydrophilic material
(vi) It is a modifiable material that can be tailored to specific applications.

They also possess some favourable biochemical properties such as non-toxicity, biocompatibility, biodegradability, polyfunctionality, high chemical reactivity, chirality, chelation and adsorption capacities.

1.5 Justification and the Theoretical Basis for the Use of Polysaccharides

1.5.1 Adsorption-based Water Treatment Operations

Activated carbon has been the conventional adsorbent in the adsorption based water treatment systems, but the use is synonymous with high cost and tedious procedure for its preparation and regeneration. Researchers are challenged to develop other adsorbents, whose operational requirements and the economy of the process of production are simple and affordable based on the stringent operational demands and the economy of the process of production of activated carbon. The choice of an adsorbent for the removal of a particular adsorbate hinges on the relationship between the surficial chemistry of the adsorbent and the hydrochemistry of the adsorbate. Polysaccharide-based adsorbents are considered as veritable options to the conventional adsorbents because of the possession of favourable surface chemistry and structural features. It has been reported that polysaccharide-based materials contain an abundance of surface polar functional groups such as aldehydes, ketones, acids and phenolics [23, 24] required to promote chemical interactions between it and the adsorbate in the aqueous phase.

In a critical review of the recent developments in polysaccharide-based materials used as adsorbents in wastewater treatment [25], the excellent adsorption behaviour of polysaccharides is mainly attributed to the following features:

(i)   The high hydrophilicity of the polymer due to hydroxyl groups of glucose units on the surface. This promotes sufficient interactions between the adsorbent and the solute in the aqueous matrix

(ii) The presence of a large number of functional groups (acetamido, primary amino and hydroxyl groups on the surface. This broadens the possible mode of interactions between the adsorbent and the adsorbate in the aqueous system

(iii) The high chemical reactivities of the surface functional groups which promotes the derivatization of the surface to produce a suitable engineered adsorbent

(iv) The flexible structure of the polymer chain. This enhances the available surface area for interaction between the adsorbate and the adsorbent.

1.5.2   Coagulation/Flocculation Treatment Operations

Despite the conventional status of metal salts and synthetic organic polymers, a retinue of shortcomings have been attributed to their continuous usage. Some of the shortcomings include

(i)   The high costs of purchase that make them unaffordable in most of the developing and underdeveloped world

(ii) The toxicity of the primary coagulant, the residual coagulant in the treated water and the coagulant by-products to humans

(iii) The production of large sludge volumes, which makes sludge handling and management a big challenge

(iv) The great influence on the pH of the product water which makes the addition of ancillary chemical(s) for pH correction a necessity

(v) The negative ecotoxicological impacts of the sludge that make the discharge and management of the sludge difficult and problematic.

Specifically, aluminum sulfate (alum), a common coagulant globally used in water and wastewater treatment, has been reported to produce large sludge volumes [26], reacts with natural alkalinity present in the water, leading to pH reduction [27, 28], and demonstrates low coagulation efficiency in cold waters [29], which makes it difficult to use in temperate regions. Furthermore, the use of alum has raised other concerns which include ecotoxicological impacts when introduced into the environment as post-treatment sludge; impacts on human health as a result of consumption of treated water; and the cost of importing these chemicals for developing communities [27]. Also, the determination of the optimum dosage for alum coagulation requires technical and scientific skills and training on the part of the operator [30]. The use of synthetic organic polyelectrolytes has also been reported to pose some environmental problems because some of the derivatives are non-biodegradable and the intermediate products of their degradation are hazardous to human health as their monomers are neurotoxic and carcinogenic [31].

Premised on the challenges associated with the conventional coagulants in water and wastewater treatment operations, the use of low cost, eco-friendly, sustainable and ubiquitous polysaccharide coagulants, as a replacement to the conventional synthetic coagulants, is now being opted for. It has been reported that to design a safe point-of-use (POU) treatment system, polysaccharide-based coagulants are considered promising alternatives because of their inherent advantages [32]. The inherent advantages include high biodegradability, non-toxicity, and non-corrosive nature, production of less voluminous sludge and does not alter the pH of the product water. It is also thought that since the precursors from which the polysaccharide-based coagulants can be derived locally, the natural plant-based coagulants are more cost-effective than the imported conventional coagulants [33].

1.6 Conclusion

In adsorption-based water treatment operations, the choice of the required operational material is based on the surface chemistry of the adsorbent, while in coagulation/flocculation operations the choice of the operational materials is based on the possession of features that enable it to operate using any of the identified coagulation mechanisms. These features include high charge density, macromolecular skeletal framework and the ability to form insoluble species. The ability of polysaccharides to act as substitutes for the conventional operational materials required for these two unit processes i.e. adsorption and coagulation/flocculation, is based on the unique inherent features that include: abundance; low cost; pervasiveness; renewable resource; stability and hydrophilicity; and its ability to be easily engineered or modified.

References

1. C.A. Zaror, Enhanced oxidation of toxic effluents using simultaneous ozonation and activated carbon treatment. J. Chem. Technol. Biotechnol. **70**, 21–28 (1997)
2. N.A. Oladoja, A critical review of the applicability of Avrami fractional kinetic equation in adsorption-based water treatment studies. Desalin. Water Treat. 1–13 (2015)
3. Y.-S. Ho, Review of second-order models for adsorption systems. J. Hazard. Mater. **B136**, 681–689 (2006)
4. J.E. Saiers, G.M. Hornberger, L. Liang, First- and second-order kinetics approaches for modeling the transport of colloidal particles in porous media. Water Resour. Res. **30**, 2499–2506 (1994)
5. M.A. McCoy, A.I. Liapis, Evaluation of kinetic-models for biospecific adsorption and its implications for finite bath and column performance. J. Chromatogr. A **548**, 25–60 (1991)
6. S.V. Mohan, N.C. Rao, J. Karthikeyan, Adsorptive removal of direct azo dye from aqueous phase onto coal based sorbents: a kinetic and mechanistic study. J. Hazard. Mater. **90**, 189–204 (2002)
7. K. Chu, M. Hashim, Modeling batch equilibrium and kinetics of copper removal by crab shell. Sep. Sci. Technol. **38**, 3927–3950 (2003)
8. D.J. O'Shannessy, D.J. Winzor, Interpretation of deviations from pseudo-first-order kinetic-behavior in the characterization of ligand binding by biosensor technology. Anal. Biochem. **236**, 275–283 (1996)
9. S. Lagergren, Zur theorie der sogenannten adsorption gelöster stoffe. K. Sven. Vetenskapsakad. Handlingar **24**, 1–39 (1898)
10. J. Zeldowitsch, Uber den mechanismus der katalytischen oxydation von CO an MnO_2. Acta Physicochim. URSS **1**, 364–449 (1934)
11. Y.S. Ho, Adsorption of heavy metals from waste streams, Ph.D. Thesis, University of Birmingham, Birmingham, U.K. (1995)
12. Y.S. Ho, G. McKay, Sorption of dye from aqueous solution by peat. Chem. Eng. J. **70**, 115–124 (1998)
13. Y.S. Ho, G. McKay, Pseudo-second order model for sorption processes. Process Biochem. **34**, 451–465 (1999)
14. Y.S. Ho, G. McKay, The kinetics of sorption of divalent metal ions onto sphagnum moss peat. Water Res. **34**, 735–742 (2000)
15. B.A. Bolto, Soluble polymers in water purification. Prog. Polym. Sci. **20**, 987–1041 (1995)
16. M.A. Yukselen, J. Gregory, The reversibility of floc breakage. Int. J. Min. Process. **73**, 251–259 (2004)
17. D.T. Ray, R. Hogg, Agglomerate breakage in polymer-flocculated suspensions. Colloid Interface Sci. **116**, 256–268 (1987)
18. A. Dabrowski, Adsorption—from theory to practice. Adv. Colloid Interface Sci. **93**, 135–224 (2001)
19. E. Worch, *Adsorption Technology in Water Treatment- Fundamentals, Processes* (Walter de Gruyter GmbH & Co. KG, Berlin/Boston, Goettingen, 2012)
20. R.M. Barrer, *Zeolites and Clay Minerals* (Academic Press, London, 1978)
21. D.W. Breck, W.G. Eversole, R.M. Milton, T.B. Read, T.L. Thomas, Crystalline zeolites. I. The properties of a new synthetic zeolite, Type A. J. Am. Chem. Soc. **78**, 5963–5972 (1956)

22. S. Kawamura, Effectiveness of natural polyelectrolytes in water treatment. J. AWWA 88–91 (1991)

23. Y.S. Ho, W.T. Chiu, C.S. Hsu, C.T. Huang, Sorption of lead ions from aqueous solution using tree fern as a sorbent. Hydrometallurgy **73**, 55–61 (2004)

24. M. Horsfall Jr., A.A. Abia, A.I. Spiff, Kinetic studies on the adsorption of Cd2+, Cu2+ and Zn2+ ions from aqueous solutions by cassava (Manihot esculenta Cranz) tuber bark waste. Bioresour. Technol. **96**(7), 782–789 (2005)

25. G. Crini, Recent developments in polysaccharide-based materials used as adsorbents in wastewater treatment. Prog. Polym. Sci. **30**, 38–70 (2005)

26. C. James, C.R. O'Melia, Considering sludge production in the selection of coagulants. J. Am. Water Works Assoc. **74**, 158–251 (1982)

27. A. Ndabigengesere, K.S. Narasiah, B.G. Talbot, Active agents and mechanism of coagulation of turbid waters using Moringa oleifera. Water Res. **29**(2), 703–710 (1995)

28. A. Ndabigengesere, K.S. Narasiah, Quality of water treated by coagulation using Moringa oleifera seeds. Water Res. **32**, 781–791 (1998)

29. J. Haaroff, J.L. Cleasby, Comparing aluminum and iron coagulants for in-line filtration of cold waters. J. Am. Water Works Assoc. **80**, 168–175 (1988)

30. WHO, *Combating Waterborne Disease at the Household Level* (WHO Press, Geneva, Switzerland, 2007)

31. C. Rudén, Acrylamide and cancer risk—expert risk assessments and the public debate. J. Food Chem. Toxicol. **42**, 335–349 (2004)

32. M. Sciban, M. Klašnja, M. Antov, B. Skrbic, Removal of water turbidity by natural coagulants obtained from chestnut and acorn. Bioresour. Technol. **100**, 6639–6643 (2009)

33. R. Sanghi, B. Bhattacharya, V. Dixit, V. Singh, Ipomoea dasysperma seed gum: an effective natural coagulant for the decolorization of textile dye solutions. J. Environ. Manage. **81**, 36–41 (2006)

Chapter 2
Mechanistic Insight into the Coagulation Efficiency of Polysaccharide-based Coagulants

2.1 Polysaccharides-A Brief Overview

Polysaccharides possess the highest industrial capacity for adsorbents that can be useful as eco-friendly materials for water treatment, due to their prevalence [1–3]. Polysaccharides are stereoregular (natural) polymers of monosaccharides (sugars) also referred to as biopolymers. They are unique raw materials because they are inexpensive and widely available in many countries in the world. They possess biological and chemical properties such as non-toxicity, biocompatibility, biodegradability, poly-functionality, high chemical reactivity, chirality, chelation and adsorption capacities. The excellent adsorption behaviour of polysaccharides is due to certain properties. These properties include (1) high hydrophilicity due to hydroxyl groups of glucose units; (2) presence of a vast number of functional groups (acetamido, primary amino, and/or hydroxyl groups); (3) high chemical reactivity of these groups; (4) flexible structure of the polymer chain [4]. The development of new products based on polysaccharides is a promising way to overcome the disadvantages of synthetic polymers and a better way of, rationally, using renewable bio-resources. Under environmental conditions, the majority of biopolymers have rather low activity [5], and the primary task in this area is therefore to create a science-based methodology for the synthesis of functional materials from polysaccharide having significant properties that can enhance their utilization on a practical scale. This is expected to open the possibility of a deeper understanding of the nature of intermolecular interactions in aqueous solutions of biopolymers. The optimal choice for a particular technology for engineering of polysaccharides can either be by chemical modification of using polymeric reagents [6] or by preparing composites [7]. A preview in literature [8–11] showed that research in the field of polysaccharide composite materials for water remediation has made some remarkable progress.

Considering biocompatibility, a polysaccharide with anticoagulant properties can increase its biocompatibility [12, 13]. Also, polysaccharides coated with

N.A. Oladoja et al., *Polysaccharides as a Green and Sustainable Resource for Water and Wastewater Treatment*, Biobased Polymers, DOI 10.1007/978-3-319-56599-6_2

magnetite nanoparticles can be used for magnetic resonance imaging of liver tumours [14]. This is because they are hydrophilic, and when administered in approved conditions, are non-toxic.

Polysaccharides have specific functional chemical groups in their structure which make it easy to engineer them through the addition of reactive and bioactive groups to produce composites. For example, hydrogels of κ-carrageenan were chemically modified by carboxymethylation of the polymer chains and further coupled to an antibody for nano delivery applications [13].

In recent times, in adsorption design, numerous approaches have been geared towards the development of adsorbents containing natural polymers. Among these are polysaccharides such as chitin [15–17]; starch [18, 19] and their derivatives [20–24]. Their use in adsorption process has increased because of their unique structure, physicochemical characteristics, chemical stability, high reactivity and excellent selectivity towards aromatic compounds and metals, resulting from the presence of chemical reactive groups (hydroxyl, acetamido or amino functions) in polymer chains [25, 26]. Besides, the increasing number of publications on adsorption of toxic compounds by these natural polymers indicate that there is a recent interest in the synthesis of new adsorbent materials from polysaccharides.

Before the advent of chemical coagulants, the use of coagulants of natural origin has been acknowledged in indigenous water purification [27]. Antediluvian civilizations in Asia and Africa have used plant extracts and derivatives as a primary coagulant for water purification [28]. This has also been proven in the Sanskrit writings in India, dating back to 400 AD [27], and the Old Testament and Roman records, dating back to 77 AD [29]. With the invasion of synthetic chemical coagulants, traditional water clarification methods using natural coagulants were jettisoned, except in rural and developing countries, where access to the synthetic chemical coagulants are substantially limited [30]. Despite the wide acceptability and the conventional status of the synthetic chemical coagulants, the downsides manifested in the 1960s, when the negative impact on human and biota was profiled [31]. The other limitations with the use of the synthetic chemical coagulants include the relatively high costs of purchase, toxicity, large sludge volumes generation and considerable alteration of the pH of the treated water [32]. Furthermore, synthetic chemical coagulants that are based on organic polymers or polyelectrolytes have been reported to pose some environmental challenges, as some of the derivatives and byproducts are non-biodegradable and the intermediate products of their degradation are hazardous to human health, as their monomer is neurotoxic and carcinogenic [33].

2.2 Polysaccharide-based Coagulants

On the strength of the coagulating abilities exhibited by some materials of biogenic ancestries as coagulants in water treatment operations, multiplicities of green bio-based materials are continually being evaluated for their coagulating properties.

These biogenic coagulants have shown several advantages over the conventional synthetic chemical coagulants. For example, the sludge volume produced from the use of the green coagulants are much lower than that generated by the use of metal salts and the natural alkalinity of the water is not consumed during the coagulation process. Consequently, the natural coagulants have exhibited potentials that conferred on them the status of workable substitutes to synthetic chemical coagulants. The green biobased coagulants are biodegradable, safe to human health and have a wider effective dosage range for coagulation-flocculation (CF) of various colloidal suspensions. Since they can be locally grown, harvested and processed, they are usually cost effective, about the imported synthetic chemical coagulants.

An overview of the polysaccharide based coagulants (PBC) that have been investigated thus far showed that they ranged from the more widely known seeds of different plant species to bone shell extracts, bark resins and exoskeleton of shellfish extracts. A review of the research summary of twenty-one (21) types of plant-based coagulants, categorized as fruit waste and others, have been provided by Choy et al. [30]. The coagulating efficiencies of these natural materials, the barriers and prospects of commercialization were highlighted. A review of fourteen (14) plant-based natural coagulants, categorized as common vegetables and legumes has also been provided by [34]. The shortcomings of the prevailing research efforts in the use of natural coagulants were discussed to provide a platform toward the necessity for further research. To ensure comprehensive anecdote of green biobased coagulants, progress in natural polymeric coagulants, for water and wastewater purifications have also been documented by [35]. Viewpoints on the promise, limitations, and the findings on the use of these biobased coagulants were also documented.

The background information on selected polysaccharides that have been investigated as primary coagulants in water and wastewater treatment are presented in Table 2.1.

Consequent upon the glowing attributes of the PBC, it is pertinent that the underlying mechanisms of the coagulation efficiencies of these low cost, eco-friendly and pervasive coagulants should be understood to enable the users to exploit the process for optimal performance. To obtain relevant information for optimum and practical conditions for CF process, using PBC, there is the need to identify the active coagulating species, which ultimately determines the underlying coagulation mechanism of the process.

2.3 Overview of Active Coagulating Species in Polysaccharide-based Coagulants

Polysaccharides are polymers whose skeletal framework consist of monosaccharides and their derivatives. The skeletal framework could be either linear or branched, and they can contain only one type of monosaccharide (homopolysaccharides), or more (heteropolysaccharides). In CF operations, the macromolecular nature of the skeletal

Table 2.1 Background information on selected polysaccharides that have been investigated as primary coagulants

S/N	Scientific names	Common names	Family name	Country of origin	References
1	*Coccinia indica*	Ivy Gourd, Scarlet Gourd, Small Gourd, Kowai Fruit, Scarlet-Fruited Gourd	*Cucurbitaceae*	Central Africa, India and Asia	[36, 37]
2	*Hibiscus esculentus*	Okra, Lady's Finger, Gumbo, Gobo	*Malvaceae*	Old World tropics (West Africa)	[38]
3	*Luffa cylindrica*	Smooth Luffa, Egyptian Luffa, Vegetable Sponge, Sponge Guard	*Cucurbitaceae*	Old World tropics; probably Asia	[36, 38]
4	*Arachis hypogaea*	Peanut, Groundnut, Monkey Nut, Pinder, Goober	*Fabaceae*	South America	[36, 39, 40]
5	*Cicer arietinum*	Dal Seeds, Chick Pea, Bengal Gram, Garbanzo Bean	*Fabaceae*	Mediterranean region	[36, 41]
6	*Dolichos biflorus*	Horsegram, Kulthi	*Fabaceae*	Old World tropics	[42]
7	*Glycine max*	Soybean, Soya Bean	*Fabaceae*	Eastern Asia	[43]
8	*Guar gum*	Guar Bean, Cluster Bean, Guaran	*Fabaceae*	India	[44]
9	*Lablab purpureus*	Hyacinth Bean, Bonavist Bean, Chink, Country Bean, Dolichos Bean	*Fabaceae*	Old World tropics	[45]
10	*Phaseolus angularis*	Azuki Bean, Adsuki Bean, Red Bean	*Fabaceae*	Unknown Exact origin	[46]
11	*Phaseolus mungo*	Urad Bean, Black Gram, Black Lentil, Black Matpe, Urd Bean	*Fabaceae*	India	[36]

(continued)

Table 2.1 (continued)

S/N	Scientific names	Common names	Family name	Country of origin	References
12	*Pisum sativum*	Green Pea, Pea, Field Pea, Garden Pea, Stringless Snowpea	*Fabaceae*	Southwestern Asia	[36]
13	*Vigna unguiculata*	Cow Pea, Black Eyed-Pea, Southern Pea, Cowgram	*Fabaceae*	Southern Africa	[36]
14	*Phaseolus vulgaris*	Common Bean	*Fabaceae*	Central or South America	[40]
15	*Cereus repandus*	Cadushi, Giant Club Cactus, Hedge Cactus, Peruvian Apple Cactus	*Cactaceae*	South America	[47]
16	*Stenocereus griseus*	Pitaya agria, Sour Pitaya	*Cactaceae*	America	[48]
17	*Opuntia ficus indica*	Prickly pears, Tuna, Nopal	*Cactaceae*	Americas	[49–51]
18	*Oryza sativa*	Rice	*Poaceae*	China	[53]

framework is considered advantageous because it provides a significant number of active sites for particle adsorption and charge neutralization.

Generally, in the use of PBC, the ensuing elevation of the organic load of the treated water, which may result in the undesired and increased microbial activities, has been identified as a snag. To obviate this challenge, the coagulating active ingredients in the PBC are isolated from the total extract and the isolated coagulating fraction is used as such. In most cases, despite the successful isolations of the coagulating active ingredients, the proper identification of this active component has either been based on conjectural efforts shrouded in controversy [35]. Often, the identification of the coagulating active component of a particular PBC is based on reports from other researchers whose study material was entirely different from the PBC under investigation. Consequently, controversy often rages on the identity of the active coagulating species of most PBCs.

The active coagulating ingredients in PBC are polymeric in nature, but the molecular compositions and the skeletal framework of these polymers may vary from one PBC to another. Thus, the difference in molecular structures and framework is expected to influence the coagulation efficiency and the underlying mechanisms of coagulation of the PBC. An overview of the active coagulating species of the group of different PBC that have been investigated as a substitute to the conventional primary coagulants in CF operations is presented below, using the facts presented in various scientific reports as a guide.

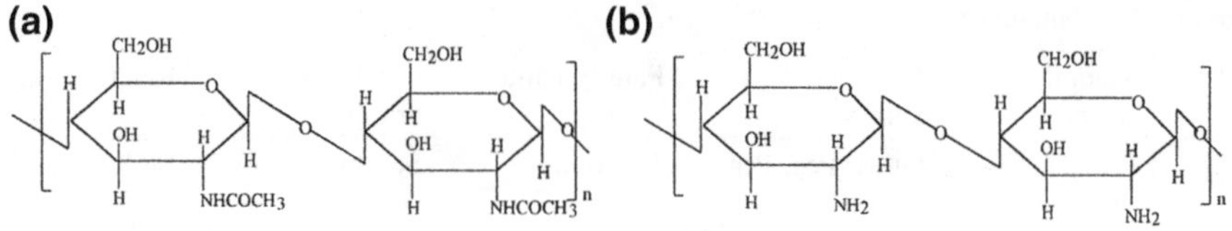

(a)   **(b)**

Fig. 2.1 Molecular units of Chitin (**a**) and Chitosan (**b**)

2.3.1 Chitosan

Among the array of investigated PBC for CF in the water industry, chitosan, is one of the most studied and has shown much promise. Chitosan is a linear copolymer of D-glucosamine and N-acetyl-D-glucosamine, produced by the deacetylation of chitin, a natural polymer of primary importance (Fig. 2.1).

Chitosan possesses both coagulating and flocculating properties (i.e., high cationic charge density, long polymer chains, bridging of aggregates and precipitation) in neutral or alkaline pH. Its uses are justified by its non-toxicity, biodegradability [27], and unusual chelation behaviour [54]. Thus, its unique physicochemical properties render it very efficient with regards to its interactions with various contaminants; including both particulate and dissolved substances.

The observed physicochemical features of chitosan have been ascribed to the intrinsic properties of amine functional groups (i.e. acid–base properties, solubility, cationic), which makes it to be very efficient for binding metal cations in near neutral solutions [55] and for interacting with anionic solutes in acidic solutions [55]. A peep into the acid–base properties of chitosan showed that the pK_a values of the amine groups strongly depends on the deacetylation degree of chitosan and the dissociation degree of the polyelectrolyte [56]. The pK_a of amine groups are close to 6.3–6.4, for entirely dissociated chitosan (with deacetylation degree close to 90%). This means that at pH 5, or below, more than 90% of amine groups get protonated.Hence, at pH below 5, most of the amine groups are protonated, and they can attract anionic species, but above the pH 5, the reverse is the case [57].

The coagulation efficiency of chitosan, in CF operations, was ascribed to the same characteristics highlighted above for adsorption reactions (i.e. cationic charge and ability to specifically bind to certain solid phases). It can effectively destabilize and coagulate natural particulate and colloidal materials, which are negatively charged, to promote the growth of large, rapid-settling floc than can then flocculate. This is because it is a long-chain polymer with positive charges (due to the high content of amine group).

2.3.2 Seed Gums

Gums are pathological products formed following an injury to plants or the effects of unfavourable conditions, such as drought, by the breakdown of cell walls.

Different seed gums have been used as eco-friendly and sustainable green PBC. It has been reported [58] that the polysaccharide composition of mature endosperm cell walls contained three mannan groups: pure mannans; glucomannans and galactomannans. An overview of the constituents of the different seed gum coagulants showed that the active coagulating component is a water-soluble, macromolecular, hydrocolloids, galactomannans, having galactose and mannose in 1:2 molar ratio (Fig. 2.2a). Galactomannans are related to mannans but contain more

(a)   α-**D-Gal**p
$$1$$
$$\downarrow$$
$$6$$
→[4-β-**D-Man**p- (1→4)-β-**D-Man**p1–] $_n$→ –

Seed gum from *Ipomoea turpethum*

α-D-Gal$p$
$$1$$
$$\downarrow$$
$$6$$
--[→ β-D-Man$p$(1→[4]-β-D-Man$p$(1→ 4)-β-D-Man$p$1--]$_n$

Seed gum from *Ipomoea quamoclit*

α-D-Gal$p$
$$1$$
$$\downarrow$$
$$6$$
--[-4)-β-D-Man$p$(1→ 4)- β-D-Man$p$1---]$_n$

Seed gum from Guar

(b)

Fig. 2.2 a Representation of the chemical structures of different Galactomannans. **b** Chemical structure of Uronic acid (Glucuronic acid)

(1,6)-β-D-galactosylside chains. These seed gums are reported [59] to consist of a linear chain of β(1→4) linked mannopyranosyl units with D-galactose side chains, attached through α(1→6) linkage to the main chain. However, they differ in their fine structure, like molecular weight and degree of polymerization.

Galactomannans with a high degree of side chains are soft and highly hydrophilic. They have been reported to be present in the endospermic leguminous seeds, such as fenugreek (*Trigonella foenum-graecum*), guar (*Cyamopsis tetragonoba*) and locust bean (*Ceratoniasiliqua*), in the endosperm of tomato (*Solanum lycopersicum*) and coffee seeds (*Coffeaarabica*), and other groups such as *Convolvulaceae*, *Asteraceae* and *Arecaceae* [58]. The presence of galactomannans has also been reported in many subfamily species *including Cassia, Senna, Leucaena, Mimosa, Prosopis, Bowdichia, Crotalaria,* and *Indigofera* [60]. Besides, the presence of galactomannans has been reported in cotyledons and hull of *Lapinus albus* [61].

The presence of an active coagulating component of the seed gum of *Sterculia lychnophora* (i.e. Malva nut seed or Taiwan sweetgum tree), whose properties are similar to uronic acids has also been reported [62]. Uronic acids (Fig. 2.2b) are a class of sugar acids with both carbonyl and carboxylic acid functional groups. They are sugars in which the terminal carbon's hydroxyl group has been oxidized to a carboxylic acid. It was posited that the high molecular weight and the presence of uronic acid by this seed gum are the requirements for the bridging and adsorption mechanism in CF operations. It was noted that the molecular weight and the intrinsic viscosity of the *S. lychnophora* seed gum were much higher than that of many polysaccharide gums (i.e., guar gum, locust bean gum, and pectin) that are currently available in the market [62].

2.3.3   Fruit Wastes

An overview of different research reports showed that extracts of an array of fruit wastes had been studied as coagulants in the treatment of turbid synthetic water, raw surface water and wastewater [30]. These fruit wastes include the seeds of *Carica papaya, Feronia limonia, Mangifera indica, Persea americana,* seeds and pollen sheath of *Phoenix dactylifera, Prunus armeniaca, Tamarindus indica,* the peels of *Citrus Sinensis* and the foliage of *Hylocereus undatus.*

The coagulating actions of these fruit wastes were ascribed mainly to the presence of proteins and polysaccharides, which are among the natural polymers present in fruit waste. The nomenclature of the particular polysaccharide that is responsible for the coagulating properties of the fruit waste is often not identified or pinpointed. The usually large molecular weights and longer polymeric chain of polysaccharides are typically regarded as the impetus for coagulation efficiency since the number of active sites that are available for particle adsorption, and subsequent flocculation abounds in such skeletal frameworks.

2.3.4 Mucilage

Mucilages are normal products of metabolism, formed within the cell (intracellular formation) and they are produced without injury to the plant. It is a thick, gluey substance produced by nearly all plants and some microorganisms. It is a polar glycoprotein and an exopolysaccharide. Mucilage in plant plays a role in the storage of water and food, seed germination, and thickening of membranes. Cacti and flax seeds are rich sources of mucilage. The mucilages of *Opuntia ficus-indica* cactus, *Hibiscusesculentus* (okra), *Plantago species*, *Malva sylvestris* (mallow) have been tested, as primary coagulants in CF operations. These plants are characterized by the production of a hydrocolloid (i.e. mucilage) which forms molecular networks that can retain significant amounts of water [63]. According to [64], the hydrocolloids are complex polymeric substances of carbohydrate nature, with a highly branched structure. It contains varying proportions of L-arabinose, D-galactose, L-rhamnose, and D-xylose, as well as the galacturonic acid in different proportions.

In *O. ficus-indica*, the mucilage structure is proposed as two distinctive water-soluble fractions. One inspection with gelling properties with Ca^{2+} and the other is mucilage, without gelling properties [65]. Majdoub et al. [66] reported that in *O. ficus-indica*, the water-soluble polysaccharide fraction with thickening properties represents less than 10% of the water-soluble material. By a working hypothesis that coagulation occurs through a polymer bridge, the polysaccharide, presenting *Opuntia* spp., are considered as the active coagulating ingredients.

In order to identify the active coagulating ingredient, in the mucilage of *opuntia* spp., individual mucilage components(D, L-arabinose, >99%; D-(+)-galactose, >99%; L-rhamnose, >99%; and D-(+)-galacturonic acid, >97%), were tested independently and in combination [67]. It was observed that the galacturonic acid component of the mucilage might be responsible for some of the turbidity reduction by *Opuntia* spp. Galacturonic acid, added independently, was able to reduce turbidity by more than 50%. Regardless, arabinose, galactose, and rhamnose displayed no coagulation activity, however, when added in combination with galacturonic acid, they reduced turbidity between 30 and 50%. The individual mucilage components, in isolation and combination, could only account for 50% of the turbidity removal observed when the full cactus pad was introduced to the turbid water solution. Consequently, it was concluded that there are additional components of the *Opuntia* spp., beyond those found in the mucilage, contributing to the observed coagulation activity. Further studies were recommended to determine the other components of the *Opuntia* spp. plant contributing to coagulation.

2.3.5 Plant Seed Extracts

The coagulation efficiency of natural coagulants, derived from the seeds of Nirmali and maize [68] mesquite bean and *Cactus latifaria* [47], *Cassia angustifolia* [69] and different seeds of leguminose species [70] have been reported. Horse chestnut (*Aesculus hyppocastanum*) from family *Sapindaceae*, and Common oak (*Quercus robur*), Turkey oak (*Quercus cerris*), Northern red oak (*Quercus rubra*) and European chestnut (*Castanea sativa*), from family *Fagaceae*, have also been studied as natural coagulants [71] in CF operations. Amongst the different seeds that have been investigated as PBC, the seed of *Moringa Oleifera* (MO) has received the greatest attention and the coagulation efficiency is often used as a baseline for the evaluation of the other PBC. On the assumption that the active coagulating species in MO is a protein, it is assumed that the active coagulating species in other seed coagulants are also proteins.

The coagulant components of the water extracts of MO has been described as a water-soluble protein with a net positive charge [72] and as dimeric cationic proteins with molecular mass of 12–14 kDa and isoelectric point (pI) values that ranged between 10 and 11 [73]. It has also been posited that the coagulating MO extract possesses a molecular mass of 6.5 kDa and a pI value greater than 10 [74]. On the other hand, it was reported that the active component from an aqueous salt extract was not a protein, polysaccharide or lipid, but an organic polyelectrolyte with a molecular weight of about 3.0 kDa [75]. It could be assumed that the nature of the water and salt extract of the coagulating species in MO seed may be different based on the different reports. In another study, the active coagulating components of MO was reported to be soluble cationic proteins and peptides, with a molecular weight ranging from 6 to 16 kDa and isoelectric pH values around 10 [74]. One of these peptides, named MO2.1, has been purified, sequenced and proven to exhibit coagulating activity on a glass powder suspension [74], bacteria and clay [76]. A non-protein component, with a molecular weight of 3 kDa has also exhibited coagulating activity in kaolin suspension [77, 78]. The isolation of a large molecular mass protein fraction, approximately 66 kDa, that exhibit coagulation activity has also been reported [79]. Consequent upon the different opinion on the nature and properties of the coagulant protein from *M. Oleifera*, more studies are being carried out to unravel the nature of the actual coagulating species in the MO seed extracts.

2.3.6 Polyphenolics

Complex polysaccharide tannin derivatives have been used in potable water and industrial effluent treatment applications [80]. The use of tannin as either the primary coagulant or coagulant aid for water treatment have been reported [80–85]. Tannins are mostly vegetal water-soluble polyphenolic compounds with a molecular weight that ranged between 500 and some thousand Daltons. Polyphenols are a structural class of mainly natural, but also synthetic or semisynthetic organic chemicals characterized by the presence of multiples of phenol in the structural units. The number and characteristics of the phenol structures underlie the unique physical, chemical, and biological (metabolic, toxic, therapeutic, etc.) properties.

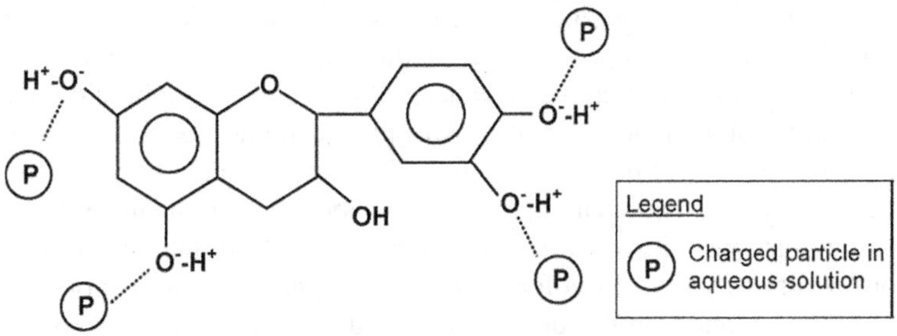

Fig. 2.3 Schematic representation of basic tannin structure in aqueous solution and possible molecular interactions [84]

The presence of phenolic groups in tannin clearly indicates its anionic nature, since it is an excellent hydrogen donor. The schematic representation of primary tannin structure, in aqueous solution, and possible molecular interactions that induce coagulation are presented in Fig. 2.3 [86].

Phenolic groups get deprotonated readily to form phenoxide, which is stabilized via resonance. This deprotonation is attributed to the delocalization of electrons within the aromatic ring, which increases the electron density of the oxygen atom. This provides an indication that the more the phenolic groups that are available in a tannin structure, the more efficient is its coagulation capability. An interesting study on the application of a commercial tanning containing both amine and phenolic groups for water treatment [87] showed that this tannin is cationic in nature since there is a single tertiary amine group per monomer, giving a charge density of approximately 3 meq/g. This tannin also exhibits amphoteric nature because of the presence of phenolic groups.

2.3.7 Starch

Starch and starch rich materials (e.g. cereals) have also been studied as primary coagulants [88, 89] and coagulant aid [90] in CF operations. In the crude form, starch consists of a mixture of two polymers of anhydroglucose units, amylose and amylopectin [91]. Amylose is a linear polymer of 1–4 linked α-D-glucopyranosyl units with low molecular weight, which makes up 25% of the starch while the rest is amylopectin, a highly branched polymer of α-D-glucopyranosyl residues linked together by 1–4 linkages with 1–6 bonds at the branch points. As a major component for most starches, amylopectin plays a critical role in defining the characteristics of the starch.

Albeit, different reports have been presented on the coagulating effects of starch in water treatment, but [35] opined that the poor cationic charge density of starch couldnot make it an efficient primary coagulant. The poor coagulating ability of starch and the constituents (i.e. amylose and amylopectin) has also been ascertained by Rath and Singh [92]. The flocculation characteristics of grafted and ungrafted starch, amylose, and amylopectin were studied, and it was observed that the behaviour of amylose could not be considered because of its insolubility in water. In cases of starch and amylopectin, there was practically no floc formation, which may be the reason for the poor flocculation efficiency exhibited [92].

Consequent upon the different reports on the poor coagulation efficiencies of the starch-based materials, it could be concluded that adsorption (a surface phenomenon) was possibly mistaken for coagulation (a phase transformation reaction) in the published reports on the use of starches derived from various sources, as primary coagulants. An overview of the applications of starches in CF operations showed that to enhance the surface charge density of starch molecules, it is modified to obtain products of excellent coagulation efficiency.

2.3.8 Actinobacteria

Coagulants of microbial origin referred to as bioflocculants, are innocuous, environmentally friendly and have been documented to show flocculation efficiency that is comparable to those of conventionally used flocculants [93–95]. Microbial flocculants are mainly polysaccharide produced mostly by bacteria such as *Alcaligenes cupidus* KT-201 [96], *A. latus* B-16 [97], and *Bacillus* sp. DP-152 [98]. Axenic cultures including *Bacillus firmus* [99], *Arthrobacter* sp. Raats [100], *Enterobacter cloacae* WD7 [101], *Streptomyces* sp. Gansen and *Cellulomonas* sp. Okoh [95] *Bacillus* sp. Gilbert [102] and *Pseudoalteromonas* sp. SM9913 [103], of the extreme deep sea psychrophilic milieu, have been respectively shown to produce bioflocculants. High production cost and low yield have been attributed to the limited application of bioflocculant in water treatment operations [104].

The bioflocculants produced from *Streptomyces* sp. Gansen and *Cellulomonas* sp. Okoh were characterized as proteoglycan and glycosaminoglycan polysaccharide, respectively, and were found to be stable to extremes of pH and high temperature [105]. Chemical analyses of the purified consortium bioflocculant revealed that polysaccharides (34.4%) and proteins (18.56%) accounted for about 52.96% of the composition. Further, analysis of the polysaccharide constituent showed the presence of neutral sugars (5.7 mg), amino sugars (9.3 mg), and uronic acids (17.8 mg), out of 100 mg of the purified bioflocculant.

The determination of the component of a partially purified *E. cloacae* WD7 biopolymer showed that it was composed of neutral sugars (29.4%) and uronic acids(14.18%) as the major and minor components, respectively, with a little amount of amino sugar (0.93%) [101]. Neither alpha amino acids, analysed by the ninhydrin reaction (L-leucine standard), nor aromatic amino acids, analysed by the Xanthoproteic reaction using L-tryptophan standard [106] were detected. This showed that it contained no amino acids or protein in its molecule, hence, the biopolymer produced by *E. cloacae* WD7 was classified as a polysaccharide.

The uronic acid contained in its molecular structure might be glucuronic acid, or galacturonic acid is generally found in the acidic polysaccharides [107].

The determination of the electric charge on the biopolymer was carried out via the addition of cetylpyridinium chloride (CPC) to the solution of partially purified biopolymer of *E. cloacae* WD7. The formation of precipitate indicated that it contained acidic groups in its structure due to the interaction with the quaternary ammonium cation (QN^+) of the CPC, resulting in the formation of a cetyl pyridinium chloride polysaccharide complex [108]. Therefore, this polymer was classified as an acidic polysaccharide; its component acid can be one or more of the acidic groups of pyruvate, succinate, uronate, acetate or sulphate [109, 110]. Margaritis et al. [107] surmised that these acidic groups may be responsible for the anionic (or acidic) charge of the polysaccharide.

2.3.9   Alginate

Alginate (chemical formula: $(C_6H_8O_6)_n$), is the term usually used to refer to the salts of alginic acid, and alginic acid itself. It is an anionic polysaccharide that binds with water to form a viscous gum. Alginates occur both as a structural component in marine brown algae (*Phaeophyceae*), comprising up to 40% of dry matter, and as capsular polysaccharides in soil bacteria. Alginates from different species of brown seaweed show variations in their chemical structure and physical properties. Alginic acid is the only polysaccharide, which naturally contains carboxyl groups in each constituent residue, and possesses various abilities for functional materials [111]. It is a linear copolymer with homopolymeric blocks of (1-4)-linked β-D-mannuronate (M) and its C-5 epimer α-L-guluronate (G) residues, respectively, covalently linked together in different sequences or blocks. The monomers can appear in homopolymeric blocks of consecutive G-residues (G-blocks), consecutive M-residues (M-blocks) or alternating M and G-residues (MG-blocks). Therefore, the knowledge of the monomeric composition is not sufficient to determine the sequential structure. Haug and Larsen [112] suggested that a second-order Markov model would be required for a general approximate description of the monomer sequence in alginates. The main difference, at the molecular level, between algal and bacterial alginates, is the presence of O-acetyl groups at C_2 (carbon in the second position) and C_3 (carbon in the third position) in the bacterial alginates [112].

The most useful and unique property of alginates is their ability to react with polyvalent metal cations, especially calcium ions to produce stable gels or insoluble polymers [113, 114]. The mechanism of coagulating action of alginate is assumed to be guided by either charge neutralization, along with bridging the gap between the particles, or by the formation of calcium alginate gel, which is especially more efficient at high calcium concentrations [115]. Calcium alginate gel combines with particles and captures (i.e. sweep coagulation mechanism) them at the stage of gel formation or after gel formation. Finally, floc formed by the gel and the particle gets dense enough and settles down [116].

2.4 Underlying Mechanisms of Coagulation-Flocculation Process

On the strength of the classifications proposed by [117], the coagulation or destabilization of suspended colloidal particles in qua system is postulated to be achieved via four mechanisms, namely the double-layer compression, charge neutralization, bridging as well as sweep coagulation. The occurrence of this mechanism of coagulation is a function of the type of coagulant used and the nature of the water matrix on which coagulation is to be performed. Any of these modes of coagulation reaction can occur singly or in combination. The synopsis of the underlying principle of each mechanism of CF is presented below:

2.4.1 Double-Layer Compression

This coagulation mechanism rely on the action of an excess electrolyte (a highly charged ionic species), added as a coagulant to the aqua system. The coagulant alters the overall ionic concentration of the system and the electrical double layer, surrounding the particulate, is compressed to the extent that the repulsive energy barrier between the particulates is lowered. This phenomenon promotes molecular attraction and subsequent micro and macro flocs formation. However, the effectiveness of this coagulation mechanism is questionable and is usually not preferred. The presence of bivalent ions (e.g. Ca^{2+} and Mg^{2+}) in water has been reported to induce some form of coagulation activities via the double-layer compression mechanism [118].

2.4.2 Charge Neutralization

This involves the adsorption of oppositely charged ionic species, present in the coagulant, on the colloidal surface. Under normal surface water conditions, colloidal particles are usually negatively charged, thus, positively charged coagulants are attracted to the colloids to induce surface charge neutralizations. The effectiveness of this mechanism is strongly dependent on the coagulant dosage introduced because particle stabilization could easily occur once the optimum dosage is exceeded.

2.4.3 Adsorption and Bridging

The bridging of particles occurs with the introduction of long-chain polymers or polyelectrolytes, as the coagulants. The coagulants are capable of extending into the solution to capture and bind multiple particulates together.

2.4.4 Sweep Coagulation

Sweep coagulation occurs through the precipitation of metal hydroxide, and the colloidal particles got enmeshed in these precipitates. Sweep coagulation could result in improved coagulation for greater removal performance in comparison with charge neutralization [119]. Consequent upon the fact that higher coagulant dosage is required for this mode of coagulation mechanism to take effect, large sludge volume has generated at the end of the coagulation process.

An inference could be drawn that a specific coagulant can only operate using some but not all the modes of coagulation mechanism judging from the principle of each of the coagulation mechanism highlighted above. For example, alum coagulants (aluminium or ferric) can only operate using sweep coagulation, charge neutralization or double layer compression, singly or in combination, but cannot operate using the adsorption and bridging mechanism. In the same vein, coagulants based on polyelectrolytes cannot operate using sweep coagulation, but it can operate with adsorption and bridging, double layer compression or charge neutralization.

2.4.5 Insight into the Coagulation Mechanism of PBC

Consequent upon the dependence of the operating coagulation mechanism, in a specific CF system, on the nature of the coagulants and the water matrix, the PBC may not function via similar underlying coagulation mechanism. Thus, it is recommended that the underlying mode of CF process should be evaluated, contextually, to obtain an informed opinion.

An overview of the proposed mechanisms of CF that has been reported by different researchers (e.g. [62, 67, 120–122] etc.), when PBC is used as the primary coagulant, showed that the adsorption and bridging coagulation and charge neutralization are the mechanisms that are usually pinpointed. Using chitosan as a case study of PBC, it has been posited that the CF ability occurs through a dual mechanism. These include coagulation by charge neutralization and flocculation by bridging mechanism. Since Chitosan is a polymer with moderate to high molecular weight and it is positively charged within the pH of natural water, it can effectively coagulate natural particulate and colloidal materials, which are negatively charged, through adsorption, charge neutralization and interparticle bridging. The coagulating ability of other PBC, whose genre were delineated in sects. 2.3.1–2.3.9, have also been ascribed to this same mechanism, because of the macromolecular nature of the skeletal framework of the active coagulating species in PBC.

A comprehensive account of the process of CF mechanism, via the adsorption and bridging (Fig. 2.4) and charge neutralization mechanism, by polymeric coagulants was described by [121].

(a)   **(b)**

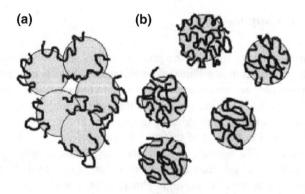

Fig. 2.4 Schematic representation of **a** an adsorption and bridging coagulation mechanism **b** restabilization by adsorbed polymeric coagulant chains [121]

It was posited that the adsorption of particulates in an aqua matrix on the polymer surface occurs only if an affinity exists between the coagulant polymer framework and the dispersed colloidal particle surface in the aqua matrix. The adsorption affinity must be sufficient to outweigh the loss of entropy associated with polymer adsorption since an adsorbed chain will have a more restricted configuration than a random coil in free solution [123]. Depending on the nature of charge on the active coagulating species in the PBC, the adsorption interaction could occur via electrostatic interaction, hydrogen bonding and ion binding. An essential requirement for bridging flocculation is that there should be sufficient unoccupied surface on a particle for attachment of segments of polymer chains adsorbed on other particles. It follows that the adsorbed amount should not be too high, else the particle surfaces will become so highly covered that there are insufficient adsorption sites available and the particles are said to be restabilised (Fig. 2.4b). It is also noteworthy that, the adsorbed amount should not be too little; otherwise not enough bridging contacts could be formed. These considerations lead to the idea of an optimum dosage bridging flocculation.

Polymer bridging gives much stronger aggregates (flocs) than those formed in other ways i.e. by metal salts [123]. This is clearly evident by the common observation that large flocs can be formed long-chain polymers even under conditions of enough high shear, as in a stirred vessel. The flocs usually grow to equilibrium (steady state) size, which is dependent on the applied shear or stirring speed. The stronger the flocs, the larger they can grow under given shear conditions [124]. Bridging contacts are also more resistant to breakage at elevated shear levels. However, floc breakage can be irreversible, so that broken flocs do not quickly re-form under reduced shear conditions [125]. Irreversible breakage may be due to scission of polymer chains under rough conditions [126] or the detachment of adsorbed polymer segments followed by re-adsorption in a manner less favourable for bridging interactions.

Most often, particulates and colloidal fractions in water matrix are negatively charged. Thus PBC whose active coagulating ingredient is cationic would be most effective as a coagulant. Electrostatic interaction gives high adsorption in such system, and the neutralisation of the particle surface and even charge reversal can occur. Thus, the possibility that coagulation could occur simply as a result of the reduced surface charge of the particles and hence a decreased electrical repulsion between them is very high. In CF operations, the optimum coagulation is achieved at coagulant dosages that are required to simply neutralise the particle surface charge, or to give a zeta potential close to zero [127]. PBC of high charge density are more effective, only because, for a given dosage, they deliver more charge to the particle surface. Since high charge density polymers tend to adsorb in a rather flat configuration, there is little opportunity for bridging interactions [123].

When high charge density PBC adsorb on negatively charged surfaces with a relatively low density of charged sites, another possibility arises, which has become known as the "electrostatic patch" mechanism [128, 129]. An important consequence of "patchwise" adsorption is that, as particles approach closely, there is an electrostatic attraction between positive patches and negative areas, which can give particle attachment and hence coagulation. Flocs produced in this way are not as strong as those formed by bridging, but stronger than flocs formed in the presence of metal salts or by simple charge neutralisation. Re-flocculation after floc breakage occurs more readily in the case of an electrostatic patch than bridging [125].

2.5   Conclusion

The active coagulating species in PBC varied widely, and the underlying mechanisms of coagulation is a function of the skeletal features of the PBC and the water matrix composition. Adsorption, charge neutralization and bridging mechanism are the coagulation-flocculation mechanism that is common in this class of coagulants.

References

1. I. Simkovic, Review: What could be greener than composites made from polysaccharides? Carbohydr. Polym. **74**, 759–762 (2008)
2. A. Tiwari, *Polysaccharides: Development, Properties and Applications* (Nova Science Publisher Inc., New York, 2010)
3. E.V. Datskevich, V.V. Goncharuk, Perspectives for the use of polysaccharides in water treatment: a short review with examples. Appl. Res. Polysaccharides, 41–71 (2015)
4. M. Hossain Md, I.H. Mondal Md, Biodegradable surfactant from natural starch for the reduction of environmental pollution and safety for water living organism. Int. J. Innov. Res. Adv. Eng. **1**, 424–433 (2014)
5. E.E. Haslan, *Comprehensive Organic Chemistry: The Synthesis and Reactions of Organic Compounds, Biological Compounds*, vol. 5 (Pergamon Press, Oxford, 1985)
6. J.F. Kennedy, Chemically reactive derivatives of polysaccharides. Adv. Carbohydr. Chem. Biochem. **29**, 305–405 (1974)
7. K. Lee, N. Morad, T. Teng, B.J. Poh, Development, characterization and the application of hybrid materials in coagulation/flocculation of wastewater: a review. Chem. Eng. J. **203**, 370–386 (2012)

8. G. Crini, Recent developments in polysaccharide-based materials used as adsorbents in wastewater treatment. Prog. Polym. Sci. **30**, 38–70 (2005)
9. L.S. Oliveira, A.S. Franca, *Food Sciences and Technology*, vol. 171 (Nova Publishers New Research, New York, 2008)
10. F. Renault, B. Sancey, P.M. Badot, G. Crini, Chitosan for coagulation/flocculation processes–an eco-friendly approach. Eur. Polym. J. **45**, 1337–1348 (2009)
11. A. Matilainen, M. Versalainen, N. Sillanpaa, Natural organic matter removal by coagulation during drinking water treatment: a review. Adv. Colloid Int. Sci. **159**, 189–197 (2010)
12. M.M. Kemp, R.J. Linhardt, Heparin based nanoparticles. WIREs Nanomed. Nanobiotechnol. **2**, 77–87 (2010)
13. T. Trindade, A.L. Daniel-Da-Silva, Biofunctional composites of polysaccharides containing inorganic nanoparticles, in *Advances in Nanocomposite Technology*, ed. by D.A. Hashim (InTech, 2011)
14. C. Corot, P. Robert, J.M. Idée, M. Port, Recent advances in iron oxide nanocrystal technology for medical imaging. Adv. Drug Deliv. Rev. **58**, 1471–1504 (2006)
15. S.E. Bailey, T.J. Olin, R.M. Bricka, D.D. Adrian, A review of potentially low-cost sorbents for heavy metals. Water Res. **33**, 2469–2479 (1999)
16. M.N.V.R. Kumar, A review of chitin and chitosan applications. React. Funct. Polym. **46**, 1–27 (2000)
17. J. Synowiecki, N.A. Al-Khateeb, Production, properties, and some new applications of chitin and its derivatives. Crit. Rev. Food Sci. Nutr. **43**, 145–171 (2003)
18. P.A. Sandford, J. Baird (eds.), *Industrial Utilization of Polysaccharides* (Academic Press, New York, 1983)
19. O.B. Wurzburg (ed.), *Modified Starches: Properties and Uses* (CRC Press, Boca Raton, 1986)
20. G. Crini, N. Morin, J.C. Rouland, L. Janus, M. Morcellet, S. Bertini, Adsorption de beta-naphtol sur des gels de cyclodextrine-carboxyme thylcellulose reticulés. Eur. Polym. J. **38**, 1095–1103 (2002)
21. M. Singh, R. Sharma, U.C. Banerjee, Biotechnological applications of cyclodextrins. Biotechnol. Adv. **20**, 341–359 (2002)
22. S. Babel, T.A. Kurniawan, Low-cost adsorbents for heavy metals uptake from contaminated water: a review. J. Hazard. Mater. **97**, 219–243 (2003)
23. A.J. Varma, S.V. Deshpande, J.F. Kennedy, Metal complexation by chitosan and its derivatives: a review. Carbohydr. Polym. **55**, 77–93 (2004)
24. E.M.M. Del-Valle, Cyclodextrins and their uses: a review. Proc. Biochem. **39**, 1033–1046 (2004)
25. E. Polaczek, F. Starzyk, K. Malenki, P. Tomasik, Inclusion complexes of starches with hydrocarbons. Carbohydr. Polym. **43**, 291–297 (2000)
26. W. Ciesielski, C.Y. Lii, M.T. Yen, P. Tomasik, Interactions of starch with salts of metals from the transition groups. Carbohydr. Polym. **51**, 47–56 (2003)
27. J. Bratby, *Coagulation and Flocculation in Water and Wastewater Treatment*, 2nd edn. (IWA Publishing, 2007)
28. M. Asrafuzzaman, A.N.M. Fakhruddin, M. Alamgir Hossain, Reduction of turbidity of water using locally available natural coagulants. ISRN Microbiol. (2011). http://dx.doi.org/10. 5402/2011/632189
29. C.C. Dorea, Use of *Moringa* spp. seeds for coagulation: a review of a sustainable option. Water Sci. Technol.: Water Supply **6**, 219–227 (2006)
30. S.Y. Choy, K.M.N. Prasad, T.Y. Wu, M.E. Raghunandan, R.N. Ramanan, Utilization of plant-based natural coagulants as future alternatives towards sustainable water clarification. J. Environ. Sci. **26**, 2178–2189 (2014)
31. G.S. Simate, S.E. Iyuke, S. Ndlovu, M. Heydenrych, L.F. Walubita, Human health effects of residual carbon nanotubes and traditional water treatment chemicals in drinking water. Environ. Int. **39**, 38–49 (2012)

32. G. Vijayaraghavan, T. Sivakumar, A. Vimal Kumar, Application of plant based coagulants for wastewater treatment. Int. J. Adv. Eng. Res. Stud. **1** (2011)
33. C. Rudén, Acrylamide and cancer risk—expert risk assessments and the public debate. Food Chem. Toxicol. **42**, 335–349 (2004)
34. D. Nkhata, Moringa as an alternative to aluminium sulphate, in Procroceedings of People and Systems for Water, Sanitation and Health 27thWEDC Conference, Lusaka, Zambia, 236–238 (2001)
35. N.A. Oladoja, Headway on natural polymeric coagulants in water and wastewater treatment operations. J. Water Process Eng. **6**, 174–192 (2015)
36. T.K. Lim, *Edible Medicinal and Non-medicinal Plants* (Springer, New York, 2012)
37. S.Z. Shaheen, K. Bolla, K. Vasu, M.A. SingaraCharya, Antimicrobial activity of the fruit extracts of *Coccinia indica*. Afr. J. Biotechnol. **8**, 7073–7076 (2009)
38. E. Small, *Top 100 Exotic Food Plants* (CRC Press, Boca Raton, 2011)
39. L. Boshou, H. Corley, *Groundnut* (CRC Press, Boca Raton, 2006)
40. N.K. Fageria, V.C. Baligar, C.A. Jones, *Growth and Mineral Nutrition of Field Crops* (CRC Press, Boca Raton, 2010)
41. F. Ahmad, P.M. Gaur, J. Croser, Chickpea (*Cicer arietinum* L.), in *Genetic Resources, Chromosome Engineering and Crop Improvement: Grain Legumes*, ed. by R.J. Singh, P. P. Jauhar (CRC Press, Boca Raton, 2005), pp. 187–217
42. M. Brink, *Macrotyloma uniflorum* (Lam.) Verde, in *Plant Resources of Tropical Africa 1. Cereals and Pulses*, ed. by M. Brink, G. Belay (PROTA Foundation/Backhuys Publishers/CTA, Wageningen, 2006)
43. R.J. Frederic, *The Book of Edible Nuts* (Dover Publications, USA, 2004)
44. R.E. Peter, W. Qi, R. Phillippa, R. Yilong, R.-M. Simon, Guargum: agricultural and botanical aspects, physicochemical and nutritional properties, and its use in the development of functional foods, in *Handbook of Dietary Fiber*, ed. by S.S. Cho, M.L. Dreher (Marcel Dekker Inc., New York, 2001)
45. E. Small, *Top 100 Food Plants* (NRC Research Press, 2009)
46. P.C.M. Jansen, *Vigna angularis* (Willd.), in *Plant Resources of Tropical Africa 1. Cereals and Pulses*, ed. by M. Brink, G. Belay (PROTA Foundation/Backhuys Publishers/CTA, Wageningen, 2006)
47. A. Diaz, N. Rincon, A. Escorihuela, N. Fernandez, E. Chacin, C. Forster, A preliminary evaluation of turbidity removal by natural coagulants indigenous to Venezuela. Process Biochem. **35**, 391–395 (1999)
48. S.L.C. Fuentes, S.I.A. Mendoza, M.A.M. López, V.M.F. Castro, M.C.J. Urdaneta, Effectiveness of a coagulant extracted from *Stenocereus griseus* (Haw.) Buxb in water purification. Rev. Téc. Ing. Univ. Zulia **34**, 48–56 (2011)
49. J.D. Zhang, F. Zhang, Y.H. Luo, H. Yang, A preliminary study on cactus as coagulant in water treatment. Process Biochem. **41**, 730–733 (2006)
50. S.M. Miller, E.J. Fugate, V.O. Craver, J.A. Smith, J.B. Zimmerman, Toward understanding the efficacy and mechanism of *Opuntia* spp. as a natural coagulant for potential application in water treatment. Environ. Sci. Technol. **42**, 4274–4279 (2008)
51. P.C. Mane, A.B. Bhosle, C.M. Jangam, S.V. Mukate, Heavy metal removal from aqueous solution by Opuntia: a natural polyelectrolyte. J. Nat. Prod. Plant Resour. **1**, 75–80 (2011)
52. B.S. Shilpaa, K. Akankshaa, P. Girish, Evaluation of cactus and hyacinth bean peels as natural coagulants. Int. J. Chem. Environ. Eng. **3**, 187–191 (2012)
53. V.B. Thakre, A.G. Bhole, Relative evaluation of a few natural coagulants. J Water Supply Res. Technol. **44**, 89–92 (1985)
54. G. Crini, P.M. Badot, Application of chitosan, a natural aminopolysaccharide, for dye removal from aqueous solutions by adsorption processes using batch studies: a review of recent literature. Prog. Polym. Sci. **33**, 399–447 (2008)
55. E. Guibal, Interactions of metal ions with chitosan-based sorvents: a review. Sep. Purif. Technol. **38**, 43–74 (2004)

56. P. Sorlier, A. Denuzière, C. Viton, A. Domard, Relation between the degree of acetylation and the electrostatic properties of chitin and chitosan. Biomacromolecules **2**, 765–772 (2001)

57. E. Guibal, J. Roussy, Coagulation and flocculation of dye-containing solutions using a biopolymer (Chitosan). React. Funct. Polym. **67**, 33–42 (2007)

58. M.S. Otegui, Endosperm cell walls: formation, composition and functions. Plant Cell Monographies **8**, 159–174 (2007)

59. V. Singh, V. Srivastava, M. Pandey, R. Sethi, R. Sanghi, *Ipomoea turpethum* seeds: a potential source of commercial gum. Carbohydr. Polym. **51**, 357–359 (2003)

60. M. Buckeridge, V.R. Panagassi, D.C. Rocha, S.M.C. Dietrich, Seed galactomannan in the classification and evolution of the Leguminosae. Phytochemistry **34**(4), 871–875 (1995)

61. A.A. Mohamed, P. Yatas-Duarte, Nonstarchy polysaccharide analysis of cotyledon and hull of *Lapinus albus*. Ceral Chem. **72**(6), 648–651 (1995)

62. Y.C. Ho, I.N. Abbas, F.M. Alkarkhi, N. Morad, New vegetal biopolymeric flocculant: a degradation and flocculation study. Iran. J. Energy Environ. **5**(1), 26–33 (2014)

63. L. Saag, G. Sanderson, P. Moyna, G. Ramos, Cactaceae mucilage composition. J. Sci. Food Agric. **26**, 993–1000 (1975)

64. B. Matsuhiro, L. Lillo, C. Saıenz, C. Urzuıa, O. Zaırate, Chemical characterization of the mucilage from fruits of *Opuntia ficus indica*. Carbohydr. Polym. **63**, 263–267 (2006)

65. F. Goycoolea, A. Caırdenas, Pectins from *Opuntia* spp.: a short review. J. Prof. Assoc. Cactus Dev. **5**, 17–29 (2004)

66. H. Majdoub, S. Roudesli, L. Picton, D. Le-Cerf, G. Muller, M. Grisel, Prickly pear nopals pectin from *Opuntia ficus indica* physicochemical study in dilute and semi-dilute solutions. Carbohydr. Polym. **46**, 69–79 (2001)

67. S.M. Miller, E.J. Fugate, V.O. Craver, J.A. Smith, J.B. Zimmerman, Toward understanding the efficacy and mechanism of *Opuntia* spp. as a natural coagulant for potential application in water treatment. Environ. Sci. Technol. **42**, 4274–4279 (2008)

68. P.K. Raghuwanshi, M. Mandloi, A.J. Sharma, H.S. Malviya, S. Chaudhari, Improving filtrate quality using agro-based materials as coagulant aid. Water Qual. Res. J. Can. **37**, 745–756 (2002)

69. R. Sanghi, B. Bhatttacharya, V. Singh, *Cassia angustifolia* seed gum as an effective natural coagulant for decolourisation of dye solutions. Green Chem. **4**, 252–254 (2002)

70. M.B. Sciban, M.T. Klasnja, J.L. Stojimirovic, Investigation of coagulation activity of natural coagulants from seeds of different leguminose. Acta Period. Technol. **36**, 81–87 (2005)

71. M. Sciban, M. Klasnja, M. Antov, B. Skrbic, Removal of water turbidity by natural coagulants obtained from chestnut and acorn. Bioresour. Technol. **100**, 6639–6643 (2009)

72. S.Y. Choy, K.M.N. Prasad, T.Y. Wu, R.N. Ramanan, A review on common vegetables and legumes aspromising plant-based natural coagulants in water clarification. Int. J. Environ. Sci. Technol. **12**(1), 367–390 (2015)

73. A. Ndabigengesere, K.S. Narasiah, B.G. Talbot, Active agents and mechanism of coagulation of turbid waters using *Moringa oleifera*. Water Res. **29**(2), 703–710 (1995)

74. U. Gassenschmidt, K.D. Jany, B. Tauscher, H. Niebergall, Isolation and characterization of a flocculating protein from *Moringa oleifera* Lam. Biochem. Biophys. Acta **1243**, 477–481 (1995)

75. T. Okuda, A.U. Baes, W. Nishijima, M. Okada, Improvement of extraction method of coagulation active components from *Moringa oleifera* seed. Water Res. **33**, 3373–3378 (1999)

76. M. Broin, C. Santaella, S. Cuine, K. Kakou, G. Peltier, T. Joet, Flocculent activity of a recombinant protein from *Moringa oleifera* Lam. seeds. Appl. Microbiol. Biotechnol. **60**, 114–119 (2002)

77. T. Okuda, A.U. Baes, W. Nishijima, M. Okada, Coagulation mechanism of salt solution-extracted active component in *Moringa oleifera* seeds. Water Res. **35**, 830–834 (2001)

78. T. Okuda, A.U. Baes, W. Nishijima, M. Okada, Isolation and characterization of coagulant extracted from *Moringa oleifera* seed by salt solution. Water Res. **35**, 405–410 (2001)
79. H. Agrawal, C. Shee, A.K. Sharma, Isolation of a 66 kDa protein with coagulation activity from seeds of *Moringa oleifera*. Res. J. Agric. Biol. Sci. **3**(5), 418–421 (2007)
80. J. Bratby, *Coagulation and Flocculation* (Uplands Press, England, 1980)
81. M. Ozacar, I.A. Sengil, Effectiveness of tannins obtained from valonia as a coagulant aid for dewatering of sludge. Water Res. **34**, 1407–1412 (2000)
82. M. Ozacar, I.A. Sengil, The use of tannins from Turkish acorns (valonia) in water treatment as a coagulant and coagulant aid. Turk. J. Eng. Environ. Sci. **26**, 255–263 (2002)
83. M. Ozacar, I.A. Sengil, Evaluation of tannin biopolymer as a coagulant aid for coagulation of colloidal particles. Colloids Surf. A **229**, 85–96 (2003)
84. N.A. Oladoja, Y.B. Alliu, A.E. Ofomaja, I.E. Unuabonah, Synchronous attenuation of metal ions and colour in aqua stream using tannin–alum synergy. Desalination **271**, 34–40 (2011)
85. J.R. Jeon, E.J. Kim, Y.M. Kim, K. Murugesan, J.H. Kim, Y.S. Chang, Use of grape seed and its natural polyphenol extracts as a natural organic coagulant for removal of cationic dyes. Chemosphere **77**, 1090–1098 (2009)
86. C.Y. Yin, Emerging usage of plant-based coagulants for water and wastewater treatment. Process Biochem. **45**, 1437–1444 (2010)
87. N. Graham, F. Gang, J. Fowler, M. Watts, Characterisation and coagulation performance of a tannin-based cationic polymer: a preliminary assessment. Colloids Surf. A **327**, 9–16 (2008)
88. I. Dogu, A.I. Arol, Separation of dark-colored minerals from feldspar by selective flocculation using starch. Powder Technol. **139**, 258–263 (2004)
89. C.Y. Teh, T.Y. Wu, J.C. Juan, Optimization of agro-industrial wastewater treatment using unmodified rice starch as a natural coagulant. Ind. Crops Prod. **56**, 17–26 (2014)
90. N.A. Oladoja, Appraisal of cassava starch as coagulant aid in the alum coagulation of congo red from aqua system. Int. J. Environ. Pollut. Solut. **2**(1), 47–58 (2014)
91. Y. Wei, F. Cheng, H. Zheng, Synthesis and flocculating properties of cationic starch derivatives. Carbohydr. Polym. **74**, 673–679 (2008)
92. S.K. Rath, R.P. Singh, Flocculation characteristics of grafted and ungrafted starch, amylose, and amylopectin. J. Appl. Polym. Sci. **66**, 1721–1729 (1997)
93. A.I. Zouboulis, X.L. Chai, I.A. Katsoyiannis, The application of the bioflocculant for the removal of humic acids from stabilized landfill leachates. J. Environ. Manag. **70**, 35–41 (2004)
94. Q. Yang, K. Luo, D. Liao, X. Li, D. Wang, X. Liu, G. Zeng, X. Li, A novel bioflocculant produced by *Klebsiella* sp. and its application to sludge dewatering. Water Environ. J. **26**, 560–566 (2012)
95. U.U. Nwodo, E. Green, L.V. Mabinya, K. Okaiyeto, K. Rumbold, L.C. Obi, A.I. Okoh, Bioflocculant production by a consortium of Streptomyces and Cellulomonas species and media optimization via surface response model. Colloids Surf. B: Biointerfaces **116**, 257–264 (2014)
96. K. Toeda, R. Kurane, Microbial flocculant from *Alcaligenes cupidus* KT 201. Agric. Biol. Chem. **55**, 2793–2799 (1991)
97. R. Kurane, Y. Nohata, Microbial flocculation of waste liquids and oil emulsions by a bioflocculant from *Alcaligenes latus*. Agric. Biol. Chem. **55**(4), 1127–1129 (1991)
98. H.H. Suh, G.S. Kwon, C.H. Lee, S.H. Kim, H.M. Oh, B.D. Yoon, Characterization of bioflocculant produced by *Bacillus* sp. DP-152. J. Ferment. Bioeng. **82**(2), 108–112 (1997)
99. H. Salehizadeh, S.A. Shojaosadati, Isolation and characterisation of a bioflocculant produced by *Bacillus firmus*. Biotechnol. Lett. **24**, 35–40 (2002)
100. L.V. Mabinya, S. Cosa, U.U. Nwodo, A.I. Okoh, Studies on bioflocculant production by *Arthrobacter* sp. Raats, a freshwater bacteria isolated from Tyume River, South Africa. Int. J. Mol. Sci. **13**, 1054–1065 (2012)

101. P. Prasertsan, W. Dermlim, H. Doelle, J.F. Kennedy, Screening, characterization and flocculating property of carbohydrate polymer from newly isolated *Enterobacter cloacae* WD7. Carbohydr. Polym. **66**, 289–297 (2006)
102. N. Piyo, S. Cosa, V.L. Mabinya, A.I. Okoh, Assessment of bioflocculant production by *Bacillus* sp. Gilbert, a marine bacterium isolated from the bottom sediment of Algoa Bay. Mar. Drugs **9**, 1232–1242 (2011)
103. W.W. Li, W.Z. Zhou, Y.Z. Zhang, J. Wang, X.B. Zhu, Flocculation behaviour and mechanism of exopolysaccharide from deep-sea psychrophilic bacterium *Pseudomonas* sp. SM9913. Bioresour. Technol. **99**, 6893–6899 (2008)
104. J. He, Q. Zhen, N. Qiu, Z. Liu, B. Wang, Z. Shao, Z. Yu, Medium Optimization for the production of a novel bioflocculant from *Halmonas* sp. V3a using response surface methodology. Bioresour. Technol. **100**, 5922–5927 (2009)
105. U.U. Nwodo, A.I. Okoh, Characterization and flocculation properties of biopolymeric flocculant (glycosaminoglycan) produced by *Cellulomonas* sp. Okoh. J. Appl. Microbiol. **114**, 1325–1337 (2012)
106. D.T. Plummer, *An Introduction to Practical Biochemistry*, 2nd edn. (McCraw-Hill, London, 1978)
107. A. Margaritis, G.W. Pace, Microbial polysaccharides, in, *Comprehensive Biotechnology*, ed. by H.W. Blanch, S. Drew, D.I.C. Wang. The Practice of Biotechnology: Current Commodity Products, vol. 3 (Pergamon Press, Oxford, 1985), pp. 1006–1040
108. J.E. Scott, *Fractionation by Precipitation with Quaternary Ammonium Salts* (Academic Press, New York, 1965)
109. G.W. Pace, R.C. Righelato (eds.), *Production of Extracellular Microbial Polysaccharides* (Springer, Berlin, 1980)
110. I.W. Sutherland, *Bacterial Exopolysaccharides—Their Nature and Production* (Academic Press, London, 1977)
111. A. Ikeda, A. Takemura, H. Ono, Preparation of low-molecular weight alginic acid by acid hydrolysis. J. Carbohydr. Polym. **42**, 421–425 (2000)
112. A. Haug, B. Larsen, Study on the composition of alginic acid by partial acid hydrolysis. Proc. Int. Sea Weed Symp. **5**, 271–277 (1966)
113. K.I. Draget, O. Smidsrod, G. Skjak-Break, Alginates from algae, in *Polysaccharides and Polyamides in the Food Industry, Properties, Products and Patents* (Wiley, Weinheim, 2005), pp. 1–30
114. A.H. King, Brown seaweed extracts (alginates), ed. by M. Glicksman (Elsevier, 1983)
115. G.T. Grant, E.R. Morris, D.A. Rees, P.J.C. Smith, D. Thom, Biological interactions between polysaccharides and divalent cations: the egg-box model. FEBS Lett. **32**, 195–198 (1973)
116. N.E. Simpson, C.L. Stabler, C.P. Simpson, A. Sambanis, The role of the $CaCl_2$–guluronic acid interaction on alginate encapsulated $\beta TC3$ cells. J. Biomater. **25**, 2603–2610 (2004)
117. J.C. Crittenden, R.R. Trussell, D.W. Hand, K.J. Howe, G. Tchobanoglous, *Water Treatment —Principles and Design*, 2nd edn. (Wiley, Hoboken, 2005)
118. J. Duan, A. Niu, D. Shi, F. Wilson, N.J.D. Graham, Factors affecting the coagulation of seawater by ferric chloride. Desalin. Water Treat. **11**, 173–183 (2009)
119. J. Duan, J. Gregory, Coagulation by hydrolysing metal salts. Adv. Colloid Interface Sci. **100–102**, 475–502 (2003)
120. J. Beltrán-Heredia, J. Sánchez-Martín, G. Frutos-Blanco, *Schinopsis balansae* tannin-based flocculant in removing sodium dodecyl benzene sulfonate. Sep. Purif. Technol. **67**, 295–303 (2009)
121. J. Beltrán-Heredia, J. Sánchez-Martín, C. Solera-Hernández, Removal of sodium dodecyl benzene sulfonate from water by means of a new tannin-based coagulant: optimisation studies through design of experiments. Chem. Eng. J. **153**, 56–61 (2009)
122. N. Chaibakhsh, N. Ahmadi, M.A. Zanjanchi, Use of *Plantago major* L. as a natural coagulant for optimized decolorization of dye-containing wastewater. Ind. Crops Prod. **61**, 169–175 (2014)

123. B. Bolto, J. Gregory, Organic polyelectrolytes in water treatment. Water Res. **41**, 2301–2324 (2007)
124. K. Muhle, Floc stability in laminar and turbulent flow, in *Coagulation and Flocculation*, ed. by B. Dobiás (Marcel Dekker, New York, 1993), pp. 355–390
125. S.Y. Yoon, Y.L. Deng, Flocculation and reflocculation of clay suspension by different polymer systems under turbulent conditions. J. Colloid Interface Sci. **278**, 139–145 (2004)
126. M.D. Sikora, R.A. Stratton, The shear stability of flocculated colloids. Tappi **64**, 97–101 (1981)
127. J. Kleimann, C. Gehin-Delval, H. Auweter, M. Borkovec, Super-stoichiometric charge neutralization in particle-polyelectrolyte systems. Langmuir **21**, 3688–3698 (2005)
128. D.R. Kasper, *Theoretical and Experimental Investigation of the Flocculation of Charged Particles in Aqueous Solution by Polyelectrolytes of Opposite Charge* (California Institute of Technology, Pasadena, 1971)
129. J. Gregory, Rates of flocculation of latex particles by cationic polymers. J. Colloid Interface Sci. **42**, 448–456 (1973)

Chapter 3
Tuning Polysaccharide Framework for Optimal Coagulation Efficiency

3.1 Introduction

Some of the conventional treatment methods for removal of suspended and colloidal materials from water include chemical coagulation, flocculation, followed by sedimentation and sand filtration. Other methods that can be used include slow sand filtration, flotation, micro-filtration and ultra-filtration. Coagulation and flocculation processes form important parts of water and wastewater treatment in the removal of suspended particles, and they can be employed in the removal of colloidal particles that impart colour to water, create turbidity, and retain bacteria and viruses.

Different types of polymers are, today, used in water treatment. They coagulate using a mechanism called *bridging* in which particles bond to their long chains and are later removed during filtration [1]. However, the use of coagulation and flocculation in water and wastewater treatment result in the formation of large amount of sludge, which proves difficult to dewater and require further treatment procedures. This complicates handling and disposal, leading to increased cost of treatment. To abate these shortcomings, synthetic polymers have gained significant attention as water treatment coagulants. Although, they do have some limitations, they exhibit a significant degree of selectivity to certain types of colloids; they form large and stable floc, but usually do not produce a clear supernatant since they are generally incapable of enmeshing all of the colloidal particles in raw water. However, their unit cost are much higher than for alum or ferric chloride; and most of them are not readily biodegradable [2].

Natural polyelectrolytes, which are extracted from certain plant and animal matter, are a prospective alternative to synthetic polymers. They are safe to human health, biodegradable and have wider effective dosage range of flocculation for various colloid suspensions [2].

© The Author(s) 2017
N.A. Oladoja et al., *Polysaccharides as a Green and Sustainable Resource for Water and Wastewater Treatment*, Biobased Polymers,
DOI 10.1007/978-3-319-56599-6_3

3.2   The Skeletal Framework of Polysaccharides

Majority of carbohydrate materials found in nature are in the form of polysaccharides. These are substances, which include not only those that are composed of glycosidically linked sugar residues, but also molecules that contain polymeric saccharide structures that are linked through covalent bonds to amino acids, peptides, proteins, lipids and so on.

Polysaccharides also called glycans; consist of monosaccharides and their derivatives. A polysaccharide, which is composed of only one kind of monosaccharide molecule, is known as a homopolysaccharide, or homoglycan, while those that contain more than one type of monosaccharide are called heteropolysaccharides or heteroglycans. The most common constituent of polysaccharides is D-glucose. However, others such as D-fructose, D-galactose, L-galactose, D-mannose, L-arabinose, and D-xylose are also known. Some monosaccharide derivatives found in polysaccharides include the amino sugars (D-glucosamine and D-galactosamine) and their derivatives (N-acetylneuraminic acid and N-acetylmuramic acid), as well as simple sugar acids (glucuronic and iduronic acids). Glucose homopolysaccharides also known as glucans and mannose homopolysaccharides also known as mannans are so named based on the type of sugar unit that makes up the homopolysaccharides. The nature of the monosaccharides, the length of the chains and the amount of chain branching are the features that make the difference among polysaccharides. Polysaccharides can form branched structures due to the presence of hydroxyl radicals that are present on their sugar units, thus making them acceptor of glycosyl substituents, a feature that distinguishes polysaccharides from nucleic acids and protein.

The two types of polysaccharides are storage and structural polysaccharides. Storage polysaccharides occur in plants such as starch in the form of α-amylose and amylopectin. Structural polysaccharides have similar compositions as storage polysaccharides. However, they have features that differ from those of the storage polysaccharides. The most abundant natural polymer, cellulose, is a structural polysaccharide. It is found in the cell walls of almost all plants and marine algae [3–5]. The second most abundant natural polymer is chitin [6] which is widely distributed in marine invertebrates, insects, fungi, and yeast [7].

Polysaccharides are stereoregular polymers of monosaccharides (sugars). In nature, they are abundant biopolymers; inexpensive (low-cost biopolymers); renewable resources; stable and hydrophilic biopolymers. They also have biological and chemical properties such as non-toxicity, biocompatibility, biodegradability, polyfunctionality, high chemical reactivity, chirality, chelation and adsorption capacities. The excellent adsorption behavior of polysaccharides is mainly attributed to (1) high hydrophilicity of the polymer due to hydroxyl groups of glucose units; (2) presence of a vast number of functional groups (acetamido, primary amino and/or hydroxyl groups); (3) high chemical reactivity of these groups; (4) flexible structure of the polymer chain [6]. Despite the properties of polysaccharides, some of them possess undesirable properties. Chitosan and Cyclodextrins are soluble in acidic media except with physical and chemical modifications. Starch being soluble in water restricts the development of starch-based materials. However, hydrophobic materials can be produced through chemical derivatization depending on the purpose for which the material is intended. Conversion of hydroxyl groups to aminopropyl [7] or hydroxyalkyl [8] is possible when a water-soluble starch derivative is intended for application in wastewater treatment.

3.3   Justification and Theoretical Basis for Modifying Polysaccharide Framework

Chemical modification of polysaccharides is justifiable to improve their qualities regarding solubility, biocompatibility, biodegradability, mechanical properties and shape. Modification of Polysaccharides can be approached in several ways among which are:

- Blending or chemical linkages with synthetic biopolymers;
- Surface coating of micro- or nano-spheres with biocompatible synthetic polymers;
- Crosslinking with different physical or chemical reagents;
- Hydrophobization through alkylation reactions;
- Modulation of guluronic/mannuronic ratio;
- Modulation of deacetylation degree.

Polysaccharides are known to possess a large number of reactive groups (hydroxyl and acetamido groups) present at the 2-, 3-, and 6-positions in the glucose unit. Direct substitution reactions such as esterification or etherification reactions and chemical modifications (chemical derivatisation) such as hydrolysis, oxidation or grafting, or enzymatic degradation can occur on the groups of the polysaccharides to produce different polysaccharide derivatives for specific applications.The starch and chitin derivatives can be classified into three main classes of polymers: (i) modified polymers such as cationic starches, carboxymethylchitin, (ii) Derivatized biopolymers, including chitosan, cyclodextrins and their derivatives (iii) polysaccharide-based materials such as resins, gels, membranes, composite materials [6]. The derivatives of these polymers are versatile as they contain numerous functional groups, which are readily available, depending on the experimental conditions, for chemical reactions. In general, modified polysaccharides derivatives can be obtained at different mono, di or tri- positions; this allows for the production of more polar sorbents. Another important feature of starch and chitin is the ability to undergo chemical derivatization in which the grafting of some functional (hydrophobic) group onto to the polysaccharide network can improve their adsorption properties. The chemical modification of the starch and chitin also allows preparation of two derivatized polysaccharides, cyclodextrin and chitosan, respectively.

An important class of starch derivatives are the cyclodextrins or cycloamyloses. Cyclodextrins (CDs) are macrocyclic oligosaccharides polymers that contain six to twelve glucose units. They are formed by the action of an enzyme on starch. There are three microcyclic CDs; these are alpha-cyclodextrin, beta-cyclodextrin and gamma-cyclodextrin which consist of six, seven and eight D(C)-glucopyranose units, respectively. An important feature of cyclodextrins is their ability to form inclusion compounds with various molecules, especially aromatics: the interior cavity of the molecule provides a relatively hydrophobic environment into which an apolar pollutant can be trapped [9–12].

Chitosan, a deacetylated derivative of chitin is more important than chitin although the N-deacetylation is almost never complete. Chitosan is a linear poly-cationic polymer which contains 2-acetamido-2-deoxy-b-D-glucopyranose and 2-amino-2-deoxyb-D-glucopyranose residues [13]. Chitosan has been reported to chelate five to six times greater concentrations of metals than chitin [14]. This property is related to the free amino groups exposed in chitosan because of deacetylation of chitin [13]. This biopolymer has drawn particular attention as a complexing agent due to its low cost as compared to activated carbon and its high content of amino and hydroxy functional groups showing high potentials for a wide range of molecules, including phenolic compounds, dyes and metal ions [6, 15]. Metal complexation by chitosan and its derivatives has been reported [16]. Through chemical reactions such as crosslinking and grafting, modification of polysaccharides can give interesting macromolecular superstructures, e.g. gels and hydrogels networks, polymeric resins, beads, membranes, fibres or composite materials. These polysaccharide-based materials can then be used as adsorbents [6].

Polysaccharide-based materials can be modified in two ways:

(i) **Cross-linking reactions**: Polysaccharides can be cross-linked by a reaction between the hydroxyl or amino groups of the chains with a coupling agent to form water-insoluble cross-linked networks [9, 16–19]. Gels are mainly classified into two: chemical gels and physical gels. While chemical gels are formed by irreversible covalent links, physical gels are formed by various reversible links [20, 21]. An alternative to covalently cross-linked gels is the formation of polysaccharide gels by polyelectrolyte complexation [6].

(ii) **Coupling or grafting reaction**: Polysaccharide-based materials can be synthesized when polysaccharides are immobilized onto insoluble supports through coupling or grafting reactions. The resulting materials combine the physical and chemical properties of both inorganic and organic components [9, 18, 22–26].

A relatively convenient method of preparation of polysaccharide-based materials is through cross-linking reaction. Crosslinking occur when a cross-linking agent reacts with macromolecules linear chains (i.e. cross-linking step) and itself (i.e. polymerization step) in an alkaline medium. This reduces segment mobility in the polymer, and some chains are interconnected by the formation of new inter-chain linkages, thus forming a three-dimensional network. If the degree of reticulation is sufficiently high, the matrix of the polymers becomes insoluble in water and organic solvents, however, it becomes swollen in water. Hydrophilicity and degree of cross-linking are the most important factors controlling the extent of adsorption properties of a polysaccharide-based material [27]. Gels that are covalently cross-linked are called permanent or chemical gels. Unlike physical gels, chemicals gels are heterogeneous.

Solid state ^{13}C NMR study of cross-linked epichlorohydrin–cyclodextrin polymers was reported to contain a cross-linked epichlorohydrin–cyclodextrin composite and a polymerized epichlorohydrin chains obtained when EPI reacts with itself [9]. A classical reaction scheme is shown in Fig. 3.1.

The scheme in Fig. 3.1 is accepted for the reaction of EPI with CD (or starch). However, EPI can react with chitosan and also, with the amino groups in the

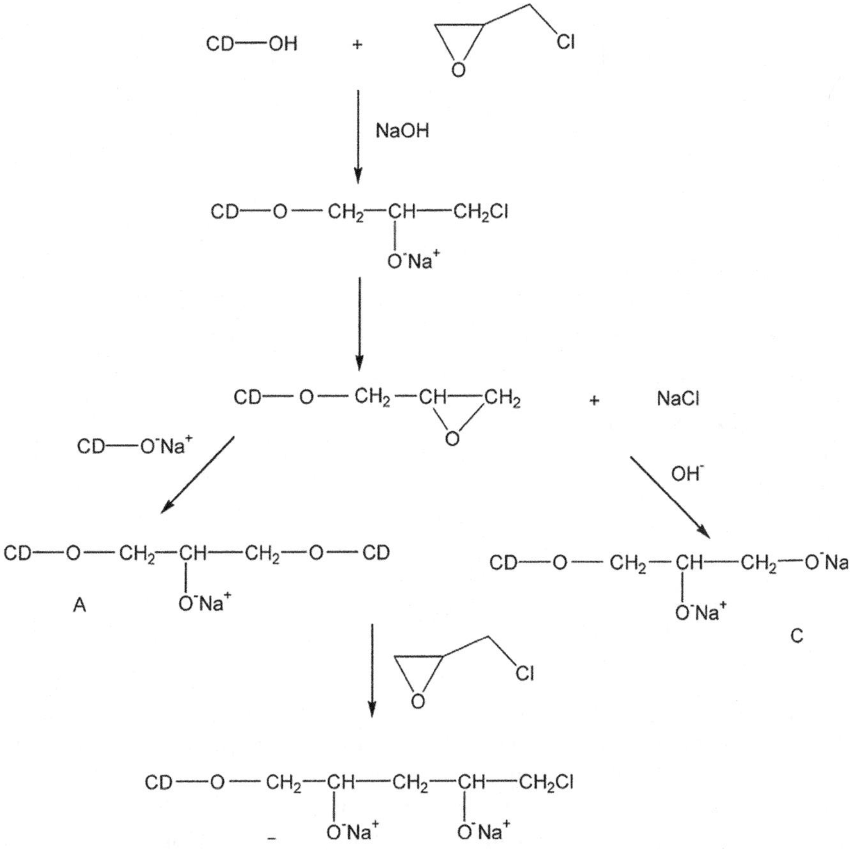

Fig. 3.1 Proposed mechanism for the reaction of EPI with cyclodextrin **a** cross-linking step, **b** cross-linked agent polymerization, **c** hydrolysis of the mono-grafted EPI leading to a glycerol monoether [8]

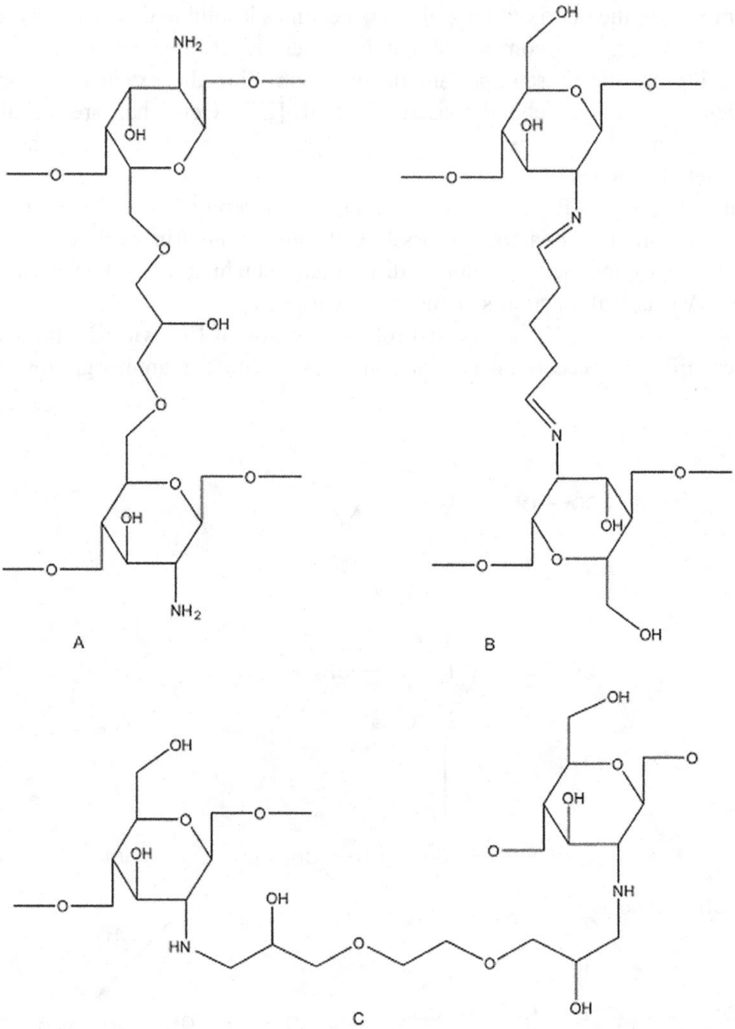

Fig. 3.2 Schematic representation of crosslinked chitosan beads: **a** chitosan–EPI, **b** chitosan–GLA, and **c** chitosan–EGDE [6]

chitosan framework resulting in more complex cross-linked chitosan beads (Fig. 3.2). Another NMR relaxation study showed that cross-linking reaction involving starch and polymer containing tertiary amine groups was not homogeneous and increasing the cross-linking degree increased the amorphous content [28]. A possible structure for cross-linked starch-polymer containing tertiary amine groups is presented in Fig. 3.3. EPI has broad industrial application. For example, it is employed as intermediates for the synthesis of many chemical products such as epoxy-resins, glycerin, polyurethane foam, elastomers, surfactants, lubricants,

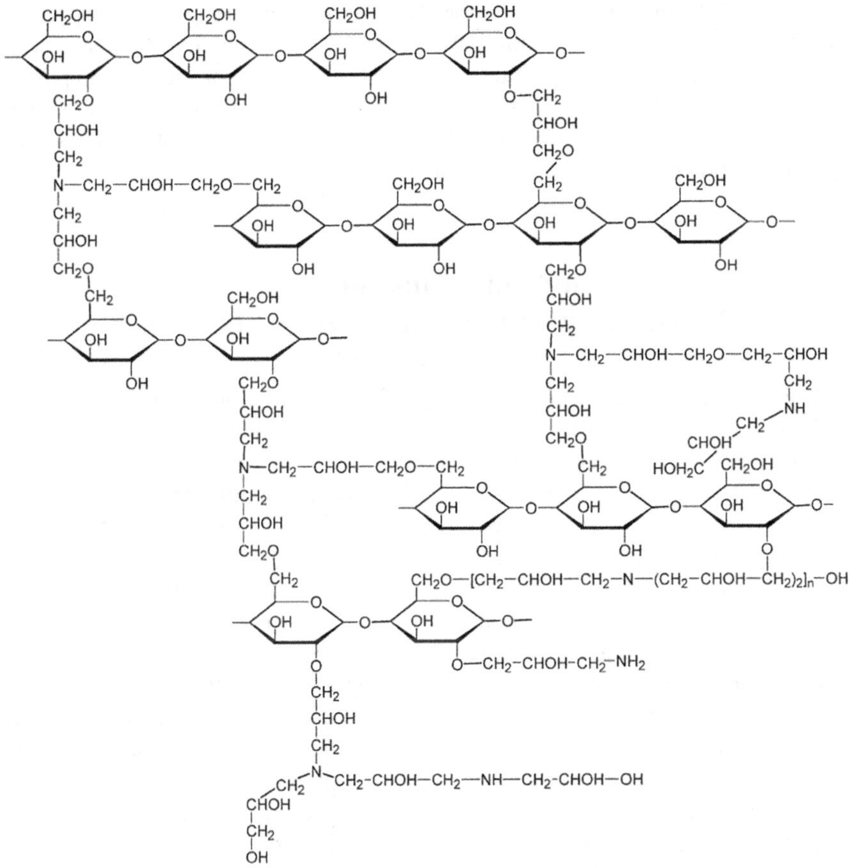

Fig. 3.3 A possible structure of a cross-linked starch-based ion exchanger [6]

drugs, etc. EPI, being a bifunctional molecule, is highly reactive with hydroxyl groups [6]. In its use in water or wastewater treatment in its modifed form with chitosan, EPI does not eliminate the cationic amine function of chitosan, which is the primary adsorption site attracting the pollutant during adsorption. However, EPI is sparingly soluble in water and partially decomposes to glycerol [6, 28]. Some Researchers are still skeptical about the cross-linking reaction between EPI and polysaccharides [9, 18], despite that the reaction has been in existence for many years now [29, 30].

Another significant cross-linking agent is glutaraldehyde (GLA), which is a dialdehyde. GLA has frequently been cross-linked with chitosan [31, 32]. The reaction is a Schiff's base reaction between aldehyde groups of GLA and some amine and hydroxyl groups of chitosan [33]. GLA is a known neurotoxin [31]. Several researchers have reported that cross-linking reactions involving polysaccharides and bi- or polyfunctional crosslinking agents such as epichlorohydrin

[33–35], ethylene glycol diglycidyl ether [22], glutaraldehyde [36, 37], caprolactone [38], lactone [39], sodium trimetaphosphate (STMP), sodium tripolyphosphate (STPP), [40] have resulted in either homogenous or heterogeneous composite material. So, STMP has been proposed as a non-toxic and an effective cross-linking agent for starch. However, some researchers have combined different modifications to produce different characteristics of starches [41–47].

3.4 Advances and Mechanistic Insight into the Modification of Polysaccharide Framework

Synthetic polymers such as polyacrylamide (PAM) and its derivatives have been commercially used as a flocculant for the removal of contaminants in water [48, 49]. However, synthetic polymers are non-biodegradable and unstable to shear field. Thus, their effective application as a flocculant is limited [50]. Conversely, natural polymers such as polysaccharides are biodegradable and resistant to shear degradation but they are poor flocculant. Grafting of PAM onto the backbone of polysaccharides gives a synergistic effect on the synthetic polymer-polysaccharide product. Flocculation efficiency is enhanced as a result of the proximity of the suspended particles to the PAM side chains.

Some studies have shown that PAM can be grafted onto the backbone of chitosan. Chitosan is a biopolymer, extracted from chitin by deacetylation in the presence of alkali [50]. Chitin is a hard, inelastic nitrogenous polysaccharide extracted from crustacean shells, such as prawns, crabs, insects, and shrimps [51, 52]. The high percentage of nitrogen content in chitosan attracts the attention of many researchers. Chitosan also contains amine and hydroxyl groups, which act as chelating sites for metal ions. It is nontoxic and biodegradable. Table 3.1 presents characteristics of chitosan preparations used in the flocculation of a bentonite suspension while Table 3.2 presents characteristics of chitosan preparations used in the flocculation of organic pollutants.

3.4.1 Modified Chitosan Flocculants

Although chitosan has been used directly as a flocculant, its shortcomings have been low molecular weight, inactive chemical properties, and poor water solubility, all of which have rendered it less efficient as a flocculant [53]. Chemical modification has been suggested to mitigate these drawbacks and improve its performance. This is achieved through the presence of abundance of free amines and hydroxyl groups on the chitosan backbone onto which various functional groups can be introduced. The products of the modification show improved characteristics such as water solubility, molecular weight ranges, charge density, and multi-functionality, allowing their use in a much wider range of applications [54].

Table 3.1 Characteristics of chitosan preparations used in the flocculation of a bentonite suspension [58]

Sample no.	MW (g/mol)	DD (%)	pH of flocculation	Ionic strength	Brief description
1	1.62×10^6– 4.70×10^6 [a]	−48 to 86	Not clear	NaClO$_4$, 0.01 mol/L	The optimal chitosan dose decreased linearly with increasing DD, but the MW was a more important determinant
2	2.79×10^4– 3.00×10^5 [a]	−69 to 100	−6.5 to 7.5	NaCl, 0.001 mol/L	MW was a more important determinant than DD
3	3–5100cP[b]	−83 to 91	−5 to 9	Not clear	DD and MW had only slight effects on the coagulation/flocculation performance
4	4.51×10^4– 3.08×10^5 [c]	−78 to 95	5 or 7	Demineralized water (DW) and tap water (TW)	Coagulation was better in TW than in DW: chitosans with a higher MW were more efficient
5	0.15×10^4– 2.33×10^5 [c]	−54.6 to 95.3	−3 to 9	DW and TW	Coagulation was better in TW than in DW; flocculation efficiency depended on not only the DD and MW but also the ionic strength, pH, and the flocculant dose

[a]Viscosity—average molecular weight
[b]Centipoise, as a direct measure of viscosity
[c]Weight—average molecular weight
MW-Molecular Weight
DD-Degree of Deacetylation

Chemical modification affords the engineering of polysaccharides depending on the nature of the pollutants and structure-activity relationships [55]. For example, cationic functional groups, such as quaternary ammonium salts, have been introduced onto the chitosan backbone to flocculate most inorganic suspended particles and many organic pollutants.

For the purpose of wastewater treatment, the two primary chemical modification methods for the modification of polysaccharide-based flocculants are etherification/amination and graft copolymerization. Both modification techniques will be briefly discussed further.

3.4.1.1 Etherification/Amination

Etherification/amination is one of the simple methods for introducing functional groups onto backbones of polysaccharides. Etherification consists of the transformation of a –OH group to form a –C–O–C– structure, yielding the etherified polymer. Amination involves the substitution of –NH$_2$ with a –C–N–C– structure,

Table 3.2 Characteristics of chitosan preparations used in the flocculation of organic pollutants [58]

Sample no.	Organic pollutant	MW (g/mol)	DD (%)	pH of flocculation	Brief description
1	Anionic dye (Reactive Black 5)	8.01×10^4, 3.08×10^5 [a]	89.5	3.5	High-MW chitosan slightly decreased the process efficiency due to the reduced availability of amine groups and the polymers less flexible structure. The molar ratio between dye molecules and amine groups ([n]) respected the stoichiometry between the sulfonic groups and the protonated amine groups at the initial pH of 5
2	Humic substances	-3.00×10^3–3.99×10^5 [b] -4.51×10^4–3.08×10^5 [a]	-51 to 99 78–95	-3.7 -4.0 to 9.0	The MW of chitosan did not significantly influence the maximum removal of humic substances. Chitosans with the highest DDs were the most efficient coagulants The MW of chitosan did not affect the coagulation/flocculation of humic acid
3	Soluble proteins in surimi wash water	-2.23×10^4–3.83×10^6 [c]	-75 to 94	5.0, 7.0	Chitosan-alginate complexes were used as flocculants at a constant mixing ratio of 0.2. The superior effectiveness of the complexes was confirmed, but differences among chitosan types did not correlate with MW and DD
4	Ink-containing packaging wastewater	8.01×10^4, 3.08×10^5 [a]	89.5	5, 7, 5	The MW had no major effects on coagulation/flocculation performance
5	Surfactant-free polystyrene latex	1.90×10^4–1.92×10^6 [c]	62–98	-3 to 7	Flocculation efficiency increased with increases in either MW or DD, but the effects were slight
6	Mushroom powder suspended in TW	-4.51×10^4–3.08×10^5 [a]	78–95	5, 7, 9	MW and DD had a limited effect on flocculation performance, which is related to pH and flocculant dose

(continued)

Table 3.2 (continued)

Sample no.	Organic pollutant	MW (g/mol)	DD (%)	pH of flocculation	Brief description
7	Organic compounds, inorganic nutrients, and bacteria in aquaculture wastewater	-3.0×10^5– 6.0×10^5 [d] -6.21×10^3– 3.62×10^5 [d]	80–98 90	-3.0 to 10.0 -3.0 to 10.0	A high DD and low pH improved the flocculation performance of chitosan High-MW chitosan was best at removing turbidity and suspended solids as well as at lowering biological and chemical oxygen demand (BOD and COD). Low-MW chitosan was best at removing NH_3 and PO_4^3 from wastewater
8	Pulp/calcium carbonate from papermaking	-7.70×10^4– 4.44×10^5 [d]	-85.7 to 87.5	Not clear	Flocculation efficiency increased with increasing MW

[a]Weigh-average molecular weigh
[b]Number-average molecular weight
[c]Viscosity-average molecular weight
[d]No details
MW-Molecular Weight
DD-Degree of Deacetylation

yielding the N-alkylated polymer. In most cases, the reactivity of $-NH_2$ is higher than that of $-OH$ [53, 56]. C_6-OH is the most reactive hydroxyl groups on the glucosamine ring of some polymer, because of the reduced steric hindrance and higher electronegativity of its oxygen. Thus, etherification reaction mostly takes place on C_6-OH groups [53, 57]. An important structural factor affecting the flocculation efficiency of a polymer-based flocculant is its degree of functional group substitution. Accordingly, during etherification/amination reactions, the dose of the modifying agent, the alkalization time, alkalization temperature, and the proportions of alkali and solvents in the reaction medium can influence the final structure of the polymer derivatives. Among them, adjusting the degree of substitution, (DS), of introduced functional groups can be easily accomplished by carefully controlling the mass to feed ratio of polymer and the modifying agent [53, 58].

Polysaccharides can also be modified through the addition of anionic groups to improve the flocculation efficiency of the polymer-based flocculant towards positively charged pollutants.

Increasing the DS within a suitable range can improve the flocculation of pollutants containing opposite surface charges, via enhanced charge neutralization effects. Furthermore, the effect of the distribution of functional groups introduced on the polymer backbone is another factor that influences flocculation [53].

3.4.1.2 Graft Copolymerization

This is another form of chemical modification of polymers in which synthetic functional polymers are introduced onto the backbones of natural polysaccharides. This allows for the formation of various molecular networks. The steps that are involved in the graft copolymerization reaction for polymers are the dissolution of the polymer into a homogenous aqueous solution; initiation of the reaction; and grafting after the feeding in of the desired amount of the selected monomer [57]. The medium for dissolution is dilute acid (e.g. ethanoic acid or hydrochloric acid), and the reaction is carried in an inert atmosphere. Furthermore, the graft reaction may be initiated by radiation (e.g. gamma ray [59]; microwave; ceric ammonium nitrate (CAN) or persulfates [53]. Examples of some of the nonionic, cationic, or anionic monomers that can be employed for the grafting reaction are listed in Table 3.3.

Copolymers synthesized through grafting reactions are characterized by their structural features, such as the grafting ratio, charge density, and the length and size of the polymer chains that are attached, which are premised on the reaction conditions, such as the total irradiation, the initiator dose, and the amount of the fed monomer [53]. The grafting ratio increases with increasing amounts of monomers [60]. The lesser the total irradiation dose or initiator concentration, the fewer but longer the side chains of graft copolymers [61].

3.4.2 Plant-based Bioflocculants

Natural plant-based bioflocculants have become a promising alternative to polymeric flocculants. Application of bioflocculants in wastewater treatment is increasingly becoming popular based on their biodegradability, nontoxicity, wide availability from renewable resources, environment friendly processing, and having no negative impact on the environment [62]. Bioflocculants derived from various plant species have been successfully employed for the treatment of biological effluent, landfill leachate, dye containing wastewater, textile wastewater, tannery effluent, and sewage effluent [63–66].

3.4.2.1 Plant Materials and Bioflocculant Preparation

Some plant species and their flocculating properties in the treatment of synthetic or real wastewater are presented in Table 3.4. The characteristic properties peculiar to these plants are the presence of mucilage and they have neutral pH. Mucilage is the hydrocolloids present in plants; they have viscous colloidal dispersion properties in water [62]. They are heterogeneous in composition and are typically polysaccharide complexes formed from the sugars of different monosaccharides, including arabinose, galactose, glucose, mannose, xylose, rhamnose, and uronic acid units [62, 65, 66].

Table 3.3 Monomers for the preparation of chitosan-based flocculants [53]

Monomer type	Monomers	Structure
Non-ionic	Acrylamide	
	N-vinyl formamide	
	N,N-dimethylacrylamide	
Cationic	(2-Methacryloyloxyethyl) trimethyl ammonium chloride	
	3-(Acrylamide)propyl trimethylammonium bromide	
	N-vinyl-2-pyrrolidone	
	diallyl dimethyl lammonium chloride	
Anionic	(Meth)acrylic acid	
	2-Acrylamidoglycolic acid	

Table 3.4 Flocculating properties of plants [62]

| Plant species | | | | | | | |
Scientific name	Common name	Charge	pH	Solubility in water	Active ingredients	Extraction method	Extracted plant part
Hibiscus/Abelmoschus Esculentus	Okra/lady fingure	Anionic	5.2–8	Soluble in cold water	L-rhamnose, D-galactose and L-galacturonic	Solvent extraction, precipitation, drying and grinding	Seedpods
Malva sylvestris	Mallow	-	6.5–7	-	-	Drying and grinding	Seedpods and lobs
Plantago psyllium	Psyllium	Anionic	7.1–7.8	Soluble in cold water	L-arabinose, D-xylose and D-galacturonic acid	Solvent extraction and precipitation	Seed husk
Plantago ovata	Isabgol	Anionic	-	-	-	Drying and grinding	Seed husk
Tamarindus indica	Tamarind	-	-	Soluble in cold water	D-galactose, D-glucose D-xylose	Solvent extraction and precipitation	Seeds
Trigonella foenum-graecum	Fenugreek	Neutral	7.7–8.6	Partially soluble	D-galactose and Dmannose	Solvent extraction	Seeds

The active ingredients in the mucilage may be responsible for the flocculating property of plant species.

Solvent extraction/precipitation and drying/grinding are the two principal methods for the production of plant-derived bioflocculants [63, 67, 68].

3.4.3 Flocculation Mechanism of Bioflocculants

Charge neutralization, electrostatic patch, and polymer bridging are the major mechanisms of flocculation via bioflocculants [69]. In some studies, X-ray diffractograms had been used to observe and postulate the possible flocculation mechanisms of natural polymers in wastewater treatment [62]. However, the X-ray diffraction patterns suggested an interaction of the solid waste with the mucilage extracted from polysaccharides. Analysis of zeta potential has been employed for the measurement of the magnitude of electrical charge around colloidal particles, since processes of coagulation and flocculation primarily depends on the electrical properties [63]. Ionic charge of bioflocculants and surface charge of suspended particles can be determined by using the measurement of zeta potential, thus, predicting flocculation mechanism. It has been reported that light diffraction scattering (LDS) can be employed to monitor the dynamics of flocculation and evaluate different types of flocculation mechanisms based on the properties of the flocculants [70, 71]. Using LDS, effects of charge density of the flocculant on the fractal dimension of the flocs, which serves as the measurement of the compactness of the aggregates can be evaluated. Monitoring the kinetics and mechanisms of the performance of a bioflocculant during flocculation process could be achieved through measurement of zeta potential and the use of LDS technique.

3.4.4 Plant-based Grafted Bioflocculants

Due to the deficiencies experienced from bioflocculants when employed alone in water and wastewater treatment processes, chemical modifications of natural macromolecules, especially polysaccharides, to improve their flocculating properties have recently gained focus [62]. Chemical modification is done to overcome deficiencies such as average flocculation efficiency, and uncontrolled biodegradability that shortens shelf life, [62]. Conversely, synthetic flocculants have higher efficiency and long shelf life but are non-biodegradable and toxic to the environment [62]. The properties of natural and synthetic polymers can be enhanced through the production of grafted copolymers [72]. The aim of graft copolymerization is to obtain a novel polymer that has a combination of the best properties of both groups of polymers involved. Grafting of synthetic polymer branches onto the backbone of natural polymer has produced efficient, reasonably shear stable and environment-friendly flocculants [62]. Synthetic polymers have been grafted onto the backbone of plant-based biopolymers to produce flocculants with enhanced flocculating properties compared to ungrafted natural polysaccharides [73–75]. The flocculating performance of the plant-based grafted bioflocculants reported in the literature is summarized in Table 3.5.

3.4.4.1 Grafting Methods

Grafted copolymers consist of a long sequence of one polymer with one or more branches of another polymer [76]. Synthesis of graft copolymer process occurs when free radical sites are created on the preformed polymer (polysaccharide) through an external agent that can create the required free radical sites onto which monomer can be added up through the chain propagation step, resulting in the formation of grafted chains. However, the agent should not rupture the structural integrity of the preformed polymer chain [76, 77]. Bioflocculants of plant origin have been successfully synthesized through conventional redox, microwave-initiated, and microwave-assisted grafting methods [62]. Figure 3.4 shows the synthesis of grafted copolymers of carboxymethyl starch (CMS), tamarind kernel polysaccharide (TKP), and sodium alginate (SAG) with acrylamide as a monomer by using a conventional, microwave-initiated, and microwave-assisted method [62].

Grafted copolymers have been successfully synthesized through conventional redox (or conventional free radical) grafting method using chemical free radical initiators such as ceric ammonium nitrate (CAN) with nitrogen as the inert gas [78–80]. Nonetheless, this grafting method has the drawback of undesired homopolymer formation in the concurrent competing reaction, which decreases the copolymer yield, contaminates the copolymeric product, and causes problems in the commercialization of the grafting procedures [59, 62]. Also, the procedure for synthesis is time-consuming as a result of prolonged Soxhlet extraction process involving removal of all the homopolymer and unreacted substrates from the copolymer surface [81]. Another drawback of the procedure is the requirement for an inert atmosphere.

To circumvent the drawbacks of the conventional redox method of grafting, synthesis of grafted copolymers using the microwave-based techniques such as a chemical free radical initiator (microwave-assisted technique) or a non-chemical free radical initiator (microwave-initiated technique) were developed [82]. These have the potential to reduce the limitations placed on the synthesis of a wide range of grafted modified polysaccharide materials. The advantages of microwave-based technique over other conventionally used methods for free radical generation are that the technique is reliable, easy to perform, and highly reproducible [62]. Grafted bioflocculants synthesized by either of the microwave techniques have produced highly efficiently grafted copolymers with a higher percentage of grafting compared to those synthesized by a conventional redox grafting method [83]. However, grafted copolymers synthesized by the microwave-assisted method

Table 3.5 Flocculating efficiencies of plant-based grafted bio-flocculants [76]

Plant-based grafted bioflocculant	Grafting method	Treated wastewater time	Sedimentation time	Optimum dose (ppm)	Type of removal	Results
Polyacrylamide-grafted-*Plantago psyllium* (Psy-g-PAN)	Conventiona free radical	Tannery and domestic wastewater	1 h	60	SS	>95% and >89% 93
		Textile wastewater	1 h	1.6	SS	>93%
					TDS	72%
					Color	15.24%
Polyacrylonitrile-grafted-*Plantago psyllium*	Conventiona free radical	Textile effluent	1 h	1.6	SS	94%
					TDS	80%
		Tannery effluent	1 h	1.2	SS	89%
					TDS	27%
Polymethacrylic acid grafted-*Plantago psyllium* (Psy-g-PMA)	Microwave assisted	Municipal sewage wastewater	25 min	2.5	Turbidity	100–12 NTU
					SS	117–14 ppm
					TDS	291–212 ppm
Polyacrylamide-grafted-*Plantago ovata*	Microwave initiated	0.25% kaolin suspension	15 min	0.75	Turbidity	59–22 NTU
		1% coal fine suspension			OD	1.5–0.25
Poly(methyl methacrylate)-grafted-*Plantago ovata*	Microwave assisted	0.25% kaolin suspension	15 min	1	Turbidity	185–70 NTU
Polyacrylamide-grafted-*Tamarindus indica* (Tam-g-PAM)	Conventiona free radical	Textile wastewater	10 min	5	Azo dye	43%
					Basic dye	27–29.6%
					Reactive dye	26.8–32.3%

(continued)

Table 3.5 (continued)

Plant-based grafted bioflocculant	Grafting method	Treated wastewater time	Sedimentation time	Optimum dose (ppm)	Type of removal	Results
Polyacrylamide-grafted-tamarind kernel polysaccharide (TKP-g-PAM)	Microwave assisted	Kaolin suspension	15 min	0.5	Turbidity	125–6 NTU
		Municipal sewage		9	Turbidity	58–14 and 97–80 NTU
		Textile industry			SS	335–55 and 295–50 ppm
		Wastewaters			TDS	265–205 and 345–295 ppm
					COD	540–205 and 586–295 ppm
Hydrolysed polyacrylamide-grafted-tamarind	Microwave	Kaolin suspension	15 min	0.5	Turbidity	125–6 NTU
Kernel polysaccharide (Hyd. TKP-g-PAM)	Assisted	Municipal sewage		9	Turbidity	58–6 NTU
					SS	335–20 ppm
					TDS	265–190 ppm
					COD	540–155 ppm

Conventional method of Synthesis of Grafted Copolymer

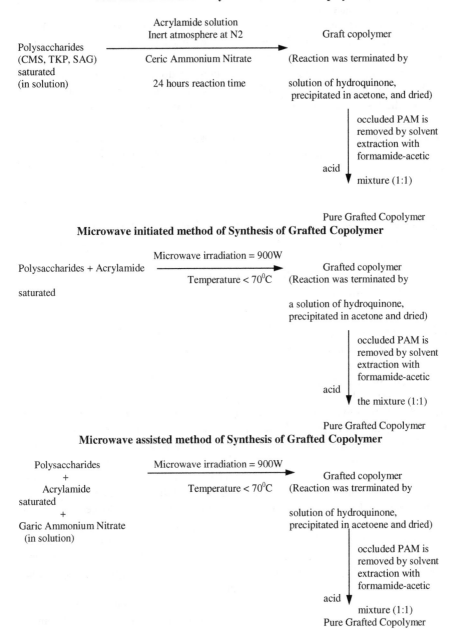

Polysaccharides
(CMS, TKP, SAG)
saturated
(in solution)

Acrylamide solution
Inert atmosphere at N2

Ceric Ammonium Nitrate

24 hours reaction time

Graft copolymer

(Reaction was terminated by

solution of hydroquinone,
precipitated in acetone, and dried)

occluded PAM is
removed by solvent
extraction with
formamide-acetic
acid
mixture (1:1)

Pure Grafted Copolymer

Microwave initiated method of Synthesis of Grafted Copolymer

Polysaccharides + Acrylamide

saturated

Microwave irradiation = 900W

Temperature < 70^{0}C

Grafted copolymer
(Reaction was terminated by

a solution of hydroquinone,
precipitated in acetone and dried)

occluded PAM is
removed by solvent
extraction with
formamide-acetic
acid
the mixture (1:1)

Pure Grafted Copolymer

Microwave assisted method of Synthesis of Grafted Copolymer

Polysaccharides
+
Acrylamide
saturated
+
Garic Ammonium Nitrate
(in solution)

Microwave irradiation = 900W

Temperature < 70^{0}C

Grafted copolymer
(Reaction was trerminated by

solution of hydroquinone,
precipitated in acetoene and dried)

occluded PAM is
removed by solvent
extraction with
formamide-acetic
acid
mixture (1:1)
Pure Grafted Copolymer

Fig. 3.4 Schematic diagram of methods of synthesis of grafted copolymers [62]

presented superior flocculation characteristics over grafted copolymers synthesized by microwave-initiated methods.

Despite the advantages displayed by grafted copolymers synthesized by microwave technology, its application for wastewater treatment is not encouraging because of the high cost of production involved. Again, most of the grafting processes reported in the literature are carried out using domestic microwave ovens in which the irradiation power is controlled by on/off cycles of the magnetron, which does not allow sufficient control of the reaction temperature and pressure hence, not safe. To subdue this shortcoming, domestic microwave ovens were modified but not without their associated high cost, which leads to high cost of production of the copolymers and the resultant limitation to their application in developing countries [62].

3.5 Performance Evaluation of Modified and Unmodified Polysaccharides

3.5.1 Performance Evaluation of Chitosan-based Flocculants

Table 3.6 presents the flocculation performance of different chitosan-based flocculants on the laboratory and pilot scale studies

Novel, amphoteric, chemically bonded composite flocculant CMC-g-PAM has been synthesized through graft copolymerization and etherification/amination reactions [68]. It has been successfully employed in the removal of an anionic (Methyl Orange) and a cationic (Basic Bright Yellow) dyes in water. It was reported that the introduction of excess PAM chains in CMC-g-PAM increased the optimal flocculant dose; decreased the colour removal efficiency, and drastically reduced the floc properties [68]. Though the PAM chains improved bridging and sweeping effects, excess amounts of it screened the charges on the chitosan backbone, thus reducing charge neutralization by CMC-g-PAM [68]. It is suggested that the grafting ratio of PAM should be controlled to a suitable range. Moreover, one or more different monomers can be used to prepare bi- or multi-graft chitosan copolymer flocculants with different functionality [58]. Series of graft chitosan flocculants have been synthesized using acrylamide and CMC as co-monomers and CAN as the initiator. Graft chitosans containing more CMC groups had a better flocculation performance, because of the increased number of positive charges and greater charge neutralization effects [58]. These chitosan flocculants exhibit long chain branches and rigid backbone, which may provide a more extended conformation of the chitosan in solution and a higher effective charge density [58, 69].

Table 3.6 Comparison of the flocculation performance of the various chitosan-based flocculants in both laboratory- and pilot-scale studies [58]

Flocculant	DS with CTA[a] (%)	DS with carboxymethyl group[a] (%)	Grafting ratio of PAM[b] (%)	Laboratory scale[c]		Pilot scale[d]	
				Optimum dose (mg/L)	TRE[e] (%)	Optimum dose (mg/L)	TRE[f] (%)
Chitosan-g-PAM	–	–	286	0.10	93.2	1.0	90.1
Chitosan-CTA-g-PAM	44.7	–	286	0.10	92.1	1.3	90.0
Chitosan-CTA	89	–	–	0.05	93.3	1.1	90.5
Carboxymethyl-Chitosan-CTA	60	48.3	–	0.80	90.4	1.1	88.4

[a]DS (%) = $W_{(substituent)}/W_{(chitosan)} \times 100\%$; here, $W_{(substituent)}$ and $W_{(chitosan)}$ are the mass weights of the substituent and chitosan, respectively

[b]Grafting ratio (%) = $W_{(grafted\ chain)}/W_{(chitosan)} \times 100\%$; here, $W_{(grafted\ chain)}$ and $W_{(chitosan)}$ are the mass weights of the grafted chain and chitosan, respectively

[c]The initial turbidity of synthetic water (kaolin suspensions) is 75 NTU (measured at 25 °C and pH 7.0)

[d]The initial turbidity of raw water (the Zhenjiang portion of the Yangtze River of China) is between 20 and 60 NTU depending on when it is measured

[e]Turbidity removal efficiency (TRE, %) = $(T_{treated} - T_{untreated})/(100 - T_{untreated}) \times 100\%$; here $T_{untreated}$ and $T_{treated}$ are the transmittances of untreated and treated water, respectively

[f]Turbidity removal efficiency (TRE, %) = $(T_{raw} - T_{treated})/T_{raw} \times 100\%$; here T_{raw} and $T_{treated}$ are the turbidity of raw water and water treated by flocculation but before sand filtration, respectively

DS-Degree of Substitution

CTA-3-chloro-2-hydroxypropyl trimethyl ammonium chloride

3.5.2   Performance Evaluation of Modified Chitosan-based Flocculants

Chitosan-g-PAM is a non-ionic grafted chitosan flocculants commonly used for wastewater treatment. Chitosan-g-PAM has been synthesized through reaction initiated by gamma ray in acid-water solution. It has also been reported that acetic acid concentration has a negligible effect on the synthesis while grafting ratio increased with increasing total irradiation dose [67]. A lower irradiation dose rate improved the grafting ratio at a fixed total irradiation dose, whereas a higher monomer concentration resulted in a higher grafting ratio [58]. Chitosan-g-PAM has higher flocculation efficiency than chitosan alone when they were employed in the flocculation of kaolin suspensions under alkaline conditions [67].

3.5.3   Performance Evaluation of Combined Flocculation with Chitosan-based Flocculants as Coagulant Aids

Chitosan-based flocculants have been applied as coagulant aids in water and wastewater treatment by combining them with other materials such as PAC, $Al_2(SO_4)_3$, $FeCl_3$, montmorillonite and soils [58, 70–73]. The synergy established through this combination has been found to improve flocculation efficiency of the combined flocculants; cost is also reduced. The applications of direct flocculation using only polymeric flocculants have been limited to organic-based wastewater with high concentration of suspended and colloidal solids; such as food, paper and pulp, and textile effluents [58, 70]. Thus, combined flocculation process is widely employed by most industries [58]. Flocculation performance of chitosan when used as a coagulant and as a coagulant aid of $Al_2(SO_4)_3$ has been investigated. The latter has higher performance over the former in the treatment of turbidity in a dam [74]. Chitosan has also been used alone and as a coagulant aid of montmorillonite in the chitosan-montmorillonite system to remove metal ions in a coagulation-flocculation process [75].

3.5.4   Performance Evaluation of Unmodified Bioflocculants In Industrial Wastewater Treatment

Solvent extraction and precipitation methods as well as drying and grinding methods were used for the preparation of bioflocculants from okra [32–37], psyllium [38], tamarind [4], and fenugreek [37, 39, 40], isabgol [28] mallow [1]. It was discovered that the bioflocculants obtained were used as a flocculant in the treatment of landfill leachate with a direct flocculation process, they had a lower removal efficiency of COD, colour, and suspended solids (SS) at 17, 27, and 41%, respectively [76]. However, the bioflocculants were found to be more effective when a coagulant was added before the addition of bioflocculant [76]. Conversely, the bioflocculants that are prepared by solvent extraction and precipitation displayed excellent flocculating ability in the treatment of wastewater with a direct flocculation process for which no coagulant and pH adjustment were required [76]. This finding indicated that the extraction step is closely related to flocculating efficiency and plays the major role in the extraction of the active constituents with high flocculating activity from the plant materials [76]. With these findings, it becomes necessary that the relationship between extraction and flocculation be investigated to evaluate the extraction parameters that may degrade the flocculating efficiency of the bioflocculants. This will allow for the optimization of the extraction conditions that favour the production of the most effective bioflocculants that are comparable to commercial flocculants regarding flocculating efficiency and cost effectiveness.

3.5.4.1   Performance Evaluation of Unmodified Bioflocculants

Applications of natural flocculants to industrial wastewater treatment are currently limited to only academic research. Some bioflocculants are effective in low concentrations and comparable to synthetic flocculants in terms of flocculation efficiency. Fenugreek and okra mucilage were proven to be as effective as a commercial flocculant (polyacrylamide) in the treatment of tannery effluent and sewage wastewater [33, 40]. As reported in the previous section, bioflocculants obtained with drying and grinding exhibit lower flocculating efficiency.

Some bioflocculants, for example, Isabgol husk, ground mallow and okra prepared by drying and grinding method were effective as a coagulant aid when grafted onto the backbones of poly(aluminium chloride) (PACl) or aluminium sulphate for the treatment of landfill leachate and removal of turbidity from kaolin suspension [1]. However, when these flocculants were prepared by solvent extraction and precipitation they displayed remarkable flocculating performance. The high removal efficiency of solids either in suspended (SS) or dissolved forms (TDS), dye, turbidity, and the colour was achieved by using a low concentration of the bioflocculant dosage. *Tamarindus indica* (Tamarind), a bioflocculant, was reported to be an effective flocculant for the removal of vat (golden yellow) and direct (direct fast scarlet) dyes from textile wastewater [76]. The suitable pH range for the maximum flocculating efficiency of the bioflocculants was neutral. However, depending on the type and characteristics of treated wastewater, some bioflocculants are effective at acidic or alkali medium.

The major hinderance to the development and application of bioflocculants in the industry are (i) sensitivity of bioproducts to preparation process, (ii) fast degradation with time, and (iii) moderate flocculating efficiency. Also, various methods of preparation of bioflocculants may affect their functional properties.

Furthermore, factors such as chemical composition and molecular structure of hydrocolloids often depend on the source and extraction methods. This is because they are susceptible to microbial attack and thus, reduced shelf-life due to high water activity and the composition of the flocculant. Also, biodegradability property of bioflocculants may cause flocs to rise, thus reducing the efficiency of the bioflocculants. Drying of biopolymers will improve the shelf-life of the biopolymer and enhance its flocculating efficiency [76].

3.5.4.2 Performance Evaluation of Modified Bioflocculants

Grafted bioflocculants of various compositions have been successfully synthesized and employed in the treatment of various types of wastewaters or effluents, leading to a reduction in environmental contaminants such as solids, turbidity, dyes, and COD. Synthetic polymers such as polyacrylamide, polyacrylonitrile, polymethacrylic acid and poly (methyl methacrylate) have been employed in the synthesis of grafted copolymers that have high flocculation efficiency than ungrafted polymers.

It has been reported that *Plantago psyllium* mucilage grafted polyacrylamide (Psy-g-PAM) copolymer exhibits higher flocculation efficiency than pure *P. psyllium* mucilage in the treatment of tannery and domestic wastewater [77]. Also, polyacrylamide grafted *Tamarindus indica* mucilage (Tam-g-PAM) exhibited higher flocculation efficiency than the pure *T. indica* mucilage for removal of various types of dyes from model textile wastewater containing azo, basic, and reactive dyes [80, 82]. In a related study, polyacrylamide grafted tamarind kernel polysaccharide (TKP-g-PAM) synthesized by microwave- assisted grafting method exhibited higher efficiency than pure TKP and polyacrylamide-based commercial flocculant (Rishfloc 226 LV) in flocculation tests [80, 82].

3.6 Conclusions

Increasing demand for environment-friendly and cleaner technologies in industries has paved way for the utilization of natural flocculants (polysaccharides) for the removal of contaminants to sustain contaminant-free environmental technology. These bioflocculants are nontoxic, biodegradable and can be obtained from renewable resources. They have displayed promising potentials as coagulants/flocculants with a high removal efficiency of solids, turbidity, colour, and dye in water. However, the development and subsequent application of the coagulants/flocculants are limited due to the variation of flocculating efficiency, fast degradation, and high production cost. Chemical modification of polysaccharides has overshadowed some the limitations and enhanced their flocculating characteristics. The synergistic effects on the synthesized modified flocculants have essentially combined the best properties of both natural and synthetic polymers and thus improved their flocculating performance. However, the shortcomings militating against modified flocculants are the complexity of the synthesis process, environmental safety issues, safety concerns of modification processes, extent of biodegradability, and high production cost.

References

1. C.P. Huang, G.S. Chen, Application of Aspergillus oryzae and Rhizopus oryzae for Cu(II) removal. Water Res. **30**(9), 1985–1990 (1996)
2. F. Cima, L. Ballarin, G. Bressa, G. Martinucci, P. Burighel, Toxicity of organotin compounds on embryos of a marine invertebrate. Ecotoxicol. Environ. Saf. **35**, 174–182 (1996)
3. M.N. Anglès, A. Dufresne, Plasticized starch/tunicin whiskers nanocomposite materials. Mechanical behaviour. Macromolecules **34**, 2921–29319 (2001)
4. P.S. Belton, S.F. Tanner, N. Cartier, H. Chanzy, High resolution solid-state 13C nuclear magnetic resonance spectroscopy of tunicin, an animal cellulose. Macromolecules **22**, 1615–1617 (1989)
5. H. Ehrlic, M. Maldonado, K.-D. Spindler, C. Eckert, T. Hanke, R. Born, C. Goebel, P. Simon, S. Heinemann, H. Worch, First evidence of chitin as a component of the skeletal fibers of marine sponges. Part I. Verongidae (Demospongia: Porifera). J. Exp. Zool. **308B**, 347–356 (2007)
6. G. Crini, Recent developments in polysaccharide-based materials used as adsorbents in wastewater treatment. Prog. Polym. Sci. **30**, 38–70 (2005)
7. A. Gonera, V. Goclik, M. Baum, P. Mischnick, Preparation and structural characterization of O-aminopropyl starch and amylose. Carbohydr. Res. **337**, 2263–2272 (2002)
8. K. Bodil-Wesslen, B. Wesslen, Synthesis of amphiphilic amylose and starch derivatives. Carbohydr. Polym. **47**, 303–311 (2002)
9. G. Crini, M. Morcellet, Synthesis and applications of adsorbents containing cyclodextrins. J. Sep. Sci. **25**, 789–813 (2002)
10. E.M.M. Del-Valle, Cyclodextrins and their uses: a review. Process Biochem. **39**, 1033–1046 (2004)
11. E. Schneiderman, A.M. Stalcup, Cyclodextrins: a versatile tool in separation science. **B745**, 83–102 (2000)
12. J. Szejtli, Introduction and general overview of cyclodextrin chemistry. Chem. Rev. **98**, 1743–1753 (1998)
13. O.S. Amuda, F.E. Adelowo, M.O. Ologunde, Kinetics and equilibrium studies of adsorption of chromium (VI) ion from industrial wastewater using Chrysophyllum albidum (Sapotaceae) seed shells. Colloids Surf., B **68**, 184–192 (2009)
14. M.F.R. Pereira, S.F. Soares, J.M.J. Orfao, J.L. Figueiredo, Adsorption of dyes on activated carbons: influence of surface chemical groups. Carbon **41**, 811–821 (2003)
15. J. Synowiecki, N.A. Al-Khateeb, Production, properties, and some new applications of chitin and its derivatives. Crit. Rev. Food Sci. Nutr. **43**, 145–171 (2003)
16. A.J. Varma, S.V. Deshpande, J.F. Kennedy, Metal complexation by chitosan and its derivatives: a review. Carbohydr. Polym. **55**, 77–93 (2004)
17. A.C. Chao, S.S. Shyu, Y.C. Lin, F.L. Mi, Enzymatic grafting of carboxyl groups on to chitosan—to confer o chitosan the property of a cationic dye adsorbent. Biores. Technol. **91**, 157–162 (2004)
18. G. Mocanu, D. Vizitiu, A. Carpov, Cyclodextrin polymers. J. Bioact. Compat. Polym. **16**, 315–342 (2001)
19. L. Janus, B. Carbonnier, A. Deratani, M. Bacquet, G. Crini, J. Laureyns, M. Morcellet, New HPLC stationary phases based on (methacryloyloxypropyl-b-cyclodextrin-co-N-

vinylpyrrolidone) copolymers coated on silica. Preparation and characterisation. New J. Chem. **27**, 307–312 (2003)

20. J. Berger, M. Reist, J.M. Mayer, O. Felt, N.A. Peppas, R. Gurny, Structure and interactions in covalently and ionically crosslinked chitosan hydrogels for biomedicals applications. Eur. J. Pharm. Biopharm. **57**, 19–34 (2004)

21. J. Berger, M. Reist, J.M. Mayer, O. Felt, N.A. Peppas, R. Gurny, Structure and interactions in chitosan hydrogels formed by complexation or aggregation for biomedical applications. Eur. J. Pharm. Biopharm. **57**, 35–52 (2004)

22. M.W. Wan, I.G. Petrisor, H.T. Lai, T.F. Yen, Copper adsorption through chitosan immobilized on sand to demonstrate the feasibility for in situ soil decontamination. Carbohydr. Polym. **55**, 249–254 (2004)

23. T. Gotoh, K. Matsushima, K.I. Kikuchi, Preparation of alginate–chitosan hybrid gel beads and adsorption of divalent metal ions. Chemosphere **55**, 135–140 (2004)

24. X. Zhang, R. Bai, Mechanisms and kinetics of humic acid adsorption onto chitosan-coated granules. J. Colloid Interface Sci. **264**, 30–38 (2003)

25. S. Jessie-Lue, S.H. Peng, Polyurethane (PU) membrane preparation with and without hydroxypropyl b-cyclodextrin and their pervaporation characteristics. J. Membr. Sci. **222**, 203–217 (2003)

26. S. Hamai, K. Kikuchi, Room-temperature phosphorescence of 6-bromo-2-naphtol in poly (vinyl alcohol) films containing cyclodextrins. J. Photochem. Photobiol., A **161**, 61–68 (2003)

27. O. Güven, M. Sen, E. Karadag, D. Saraydin, A review on the radiation synthesis of copolymeric hydrogels for adsorption and separation purposes. Radiat. Phys. Chem. **56**, 381–386 (1999)

28. D. Shiftan, F. Ravenelle, M.A. Mateescu, R.H. Marchessault, Change in the V/B polymorph ratio and T1 relaxation of epichlorohydrin crosslinked high amylose starch excipient. Starch/Stärke **52**, 186–195 (2000)

29. B.M. Gough, J.N. Pybus, Interaction of wheat starch and epichlorohydrin. Part 2. Die Stärke. **4**, 108–111 (1968)

30. L. Kuniak, R.H. Marchessault, Study of the crosslinking reaction between epichlorohydrin and starch. Die Stärke **24**, 110–116 (1972)

31. Y. Lee, W. Lee, Degradation of trichloroethylene by Fe(II) chelated with cross-linked chitosan in a modified Fenton reaction. J. Hazard. Mater. **178**(1–3), 187–193 (2010)

32. G.Z. Kyzas, D.N. Bikiaris, Recent modifications of chitosan for adsorption application: a critical and systematic review. Mar. Drugs **13**(1), 312–337 (2015)

33. M.S. Chiou, P.Y. Ho, H.Y. Li, Adsorption of anionic dyes in acid solutions using chemically cross-linked chitosan beads. Dyes Pigm. **60**, 69–84 (2004)

34. M.S. Chiou, H.Y. Li, Equilibrium and kinetic modeling of adsorption of reactive dye on cross-linked chitosan beads. J. Hazard. Mater. **B93**, 233–248 (2002)

35. F. Delval, G. Crini, J. Vebrel, M. Knorr, G. Sauvin, E. Conte, Starch-modified filters used for the removal of dyes from waste water. Macromol. Symp. **203**(1), 165–172 (2003). (WILEY-VCH Verlag)

36. M.S. Chiou, H.Y. Li, Adsorption behavior of reactive dye in aqueous solution on chemical cross-linked chitosan beads. Chemosphere **50**, 1095–1105 (2003)

37. R.S. Juang, H.J. Shao, A simplified equilibrium model for sorption of heavy metal ions from aqueous solutions on chitosan. Water Res. **36**, 2999–3008 (2002)

38. Z.Q. Shen, J. Hu, J.L. Wang, Y.X. Zhou, Comparison of polycaprolactone and starch/polycaprolactone blends as carbon source for biological denitrification. Int. J. Environ. Sci. Technol. **12**(4), 1235–1242 (2015)

39. L. Zhou, G. Zhao, W. Jiang, Mechanical properties of biodegradable polylactide/poly (ether-block-amide)/thermoplastic starch blends: effect of the crosslinking of starch. J. Appl. Polym. Sci. **133**(2), 42297 (2016)

40. F.M. Carbinatto, A.D. de Castro, B.S. Cury, A. Magalhães, R.C. Evangelista, Physical properties of pectin-high amylose starch mixtures cross-linked with sodium trimetaphosphate. Int. J. Phamaceutics **423**(2), 281–288 (2012)

41. S. Sukhcharn, S. Singh, C.S. Riar, Effect of oxidation, cross-linking and dual modification on physicochemical, crystallinity, morphological, pasting and thermal characteristics of elephant foot yam (Amorphophallus paeoniifolius) starch. Food Hydrocolloids **55**, 56–64 (2016)

42. F. Zhu, Composition, structure, physicochemical properties, and modifications of cassava starch. Carbohyd. Polym. **122**, 456–480 (2015)

43. Y. Ai, J.L. Jane, Gelatinization and rheological properties of starch. Starch-Stärke **67**, 213–224 (2015)

44. J.S. Hong, S.V. Gomand, J.A. Delcour, Preparation of cross-linked maize (Zea mays L.) starch in different reaction media. Carbohyd. Polym. **124**, 302–310 (2015)

45. Q. Chen, H. Yu, L. Wang, Z. ul-Abdin, Y. Chen, J. Wang, W. Zhou, X. Yang, R.U. Khan, H. Zhang, X. Chen, Recent progress in chemical modification of starch and its applications. R. Soc. Chem. Adv. **5**, 67459–67474 (2015)

46. P. García, C. Vergara, P. Robert, Release kinetic in yogurt from gallic acid microparticles with chemically modified inulin. J. Food Sci. **80**(10), C2147–C2152 (2015)

47. T. Ichihara, J. Fukuda, T. Takaha, S. Suzuki, Y. Yuguchi, S. Kitamura, Small-angle X-ray scattering measurements of gel produced from α-amylase-treated cassava starch granules. Food Hydrocolloids **55**, 228–234 (2016)

48. B.A. Bolto, D.R. Dixon, S.R. Gray, C. Ha, P.J. Harbour, N. Le, A.J. Ware, The use of soluble organic polymers in wastewater treatment. Water Sci. Technol. **34**(9), 117–124 (1996)

49. R. Hecker, P.D. Fawell, A. Jefferson, J.B. Farrow, Flow field-flow fractionation of polyacrylamides: commercial flocculants. Sep. Sci. Technol. **35**(4), 593–612 (2000)

50. S.K. Akbar-Ali, R.P. Singh, An Investigation of the Flocculation Characteristics of Polyacrylamide-Grafted Chitosan. J. Appl. Polym. Sci. **114**, 2410–2414 (2009)

51. J.K. Dutkiewicz, Superabsorbent materials from shellfish waste—a review. J. Biomed. Mater. Res. **63**(3), 373–381 (2002)

52. S.K. Ali, R.P. Singh, An investigation of the flocculation characteristics of polyacrylamide-grafted chitosan. J. Appl. Polym. Sci. **114**(4) (2009)

53. R. Yang, H. Li, M. Huang, H. Yang, A. Li, A review on chitosan-based flocculants and their applications in water treatment. Water Res. **95**, 59–89 (2016)

54. Z. Yang, S. Jia, N. Zhuo, W. Yang, Y. Wang, Flocculation of copper (II) and tetracycline from water using a novel pH-and temperature-responsive flocculants. Chemosphere **141**, 112–119 (2015)

55. M. Rinaudo, Chitin and chitosan: properties and applications. Prog. Polym. Sci. **31**(7), 603–632 (2006)

56. J. Lin, L. Wang, L. Wang, Coagulation of sericin protein in silk degumming wastewater using quaternized chitosan. J. Polym. Environ. **20**(3), 858–864 (2012)

57. Z. Yang, J.R. Degorce-Dumas, H. Yang, E. Guibal, A. Li, R. Cheng, Flocculation of Escherichia coli using a quaternary ammonium salt grafted carboxymethyl chitosan flocculant. Environ. Sci. Technol. **48**(12), 6867–6873 (2014)

58. Z. Yang, Y. Shang, X. Huang, Y. Chen, Y. Lu, A. Chen, Y. Jiang, W. Gu, X. Qian, H. Yang, R. Cheng, Cationic content effects of biodegradable amphoteric chitosan-based flocculants on the flocculation properties. J. Environ. Sci. **24**(8), 1378–1385 (2012)

59. V. Singh, P. Kumar, R. Sanghi, Use of microwave irradiation in the grafting modification of the polysaccharides–A review. Prog. Polym. Sci. **37**(2), 340–364 (2012)

60. B. Yuan, Y. Shang, Y. Lu, Z. Qin, Y. Jiang, A. Chen, X. Qian, G. Wang, H. Yang, R. Cheng, The flocculating properties of chitosan-graft-polyacrylamide flocculants (I)—effect of the grafting ratio. J. Appl. Polym. Sci. **117**(4), 1876–1882 (2010)

61. S.K. Ali, R.P. Singh, An investigation of the flocculation characteristics of polyacrylamide-grafted chitosan. J. Appl. Polym. Sci. **114**(4), 2410–2414 (2009)

62. C.S. Lee, M.F. Chong, J. Robinson, E. Binner, A review on development and application of plant-based bioflocculants and grafted bioflocculants. Ind. Eng. Chem. Res. **53**(48), 18357–18369 (2014)

63. Y.A. Al-Hamadani, M.S. Yusoff, M. Umar, M.J. Bashir, M.N. Adlan, Application of psyllium husk as coagulant and coagulant aid in semi-aerobic landfill leachate treatment. J. Hazard. Mater. **190**(1), 582–587 (2011)

64. K. Anastasakis, D. Kalderis, E. Diamadopoulos, Flocculation behavior of mallow and okra mucilage in treating wastewater. Desalination **249**(2), 786–791 (2009)

65. A. Mishra, M. Bajpai, Flocculation behaviour of model textile wastewater treated with a food grade polysaccharide. J. Hazard. Mater. **118**(1), 213–217 (2005)

66. A. Mishra, M. Bajpai, The flocculation performance of Tamarindus mucilage in relation to removal of vat and direct dyes. Biores. Technol. **97**(8), 1055–1059 (2006)

67. R. Srinivasan, A. Mishra, Okra (Hibiscus esculentus) and fenugreek (Trigonella foenum graceum) mucilage: characterization and application as flocculants for textile effluent treatments. Chin. J. Polym. Sci. **285**(2), 161–168 (2008)

68. A. Mishra, M. Bajpai, S. Pal, M. Agrawal, S. Pandey, Tamarindus indica mucilage and its acrylamide-grafted copolymer as flocculants for removal of dyes. Colloid Polym. Sci. **285**(2), 161–168 (2006)

69. B. Bolto, J. Gregory, Organic polyelectrolytes in water treatment. Water Res. **41**(11), 2301–2324 (2007)

70. M.G. Rasteiro, F.A.P. Garcia, P. Ferreira, A. Blanco, C. Negro, E. Antunes, The use of LDS as a tool to evaluate flocculation mechanisms. Chem. Eng. Process. **47**(8), 1323–1332 (2008)

71. M.G. Rasteiro, F.A.P. Garcia, M. del-Mar-Peréz, Applying LDS to monitor flocculation in papermaking. Part. Sci. Technol. **25**(3), 303–308 (2007)

72. S. Mishra, G.U. Rani, G. Sen, Microwave initiated synthesis and application of polyacrylic acid grafted carboxymethyl cellulose. Carbohyd. Polym. **87**(3), 2255–2262 (2012)

73. S. Ghosh, G. Sen, U. Jha, S. Pal, Novel biodegradable polymeric flocculant based on polyacrylamide-grafted tamarind kernel polysaccharide. Biores. Technol. **101**(24), 9638–9644 (2010)

74. P. Rani, G. Sen, S. Mishra, U. Jha, Microwave assisted synthesis of polyacrylamide grafted gum ghatti and its application as flocculant. Carbohyd. Polym. **89**(1), 275–281 (2012)

75. A. Mishra, R. Srinivasan, M. Bajpai, R. Dubey, Use of polyacrylamide-grafted Plantago psyllium mucilage as a flocculant for treatment of textile wastewater. Colloid Polym. Sci. **282** (7), 722–727 (2004)

76. L.Z. Liu, C. Priou, Grafting polymerization of guar and other polysaccharides by electron beams. Rhodia, Inc. U.S. Patent 7,838,667

77. S. Renneckar, State of the art paper: Biomimetics: Adapting performance and function of natural materials for biobased composites. Wood Fiber Sci. **45**(1), 3–14 (2013)

78. S. Ghosh, U. Jha, S. Pal, High performance polymeric flocculant based on hydrolyzed polyacrylamide grafted tamarind kernel polysaccharide (Hyd. TKP-g-PAM). Biores. Technol. **102**(2), 2137–2139 (2011)

79. M. Agarwal, R. Srinivasan, A. Mishra, Synthesis of Plantago psyllium mucilage grafted polyacrylamide and its flocculation efficiency in tannery and domestic wastewater. J. Polym. Res. **9**(1), 69–73 (2002)

80. R. Kumar, K. Sharma, K.P. Tiwary, G. Sen, Polymethacrylic acid grafted psyllium (Psy-g-PMA): a novel material for waste water treatment. Appl. Water Sci. **3**(1), 285–291 (2013)

81. G. Sen, S. Mishra, G.U. Rani, P. Rani, R. Prasad, Microwave initiated synthesis of polyacrylamide grafted Psyllium and its application as a flocculant. Int. J. Biol. Macromol. **50** (2), 369–375 (2012)

82. S. Mishra, A. Mukul, G. Sen, U. Jha, Microwave assisted synthesis of polyacrylamide grafted starch (St-g-PAM) and its applicability as flocculant for water treatment. Int. J. Biol. Macromol. **48**(1), 106–111 (2011)

83. S. Pal, G. Sen, S. Ghosh, R.P. Singh, High performance polymeric flocculants based on modified polysaccharides microwave-assisted synthesis. Carbohydr. Polym. **87**(1), 336–342 (2012)

Chapter 4
Progress and Prospects of Polysaccharide Composites as Adsorbents for Water and Wastewater Treatment

4.1 Progress in Preparation of Polysaccharide-based Adsorbents

There are several studies reported in many scientific literature on the preparation of adsorbents from polysaccharides for the purpose of treating water. The use of natural polysaccharides for removal of dye molecules from water have been well studied and reported [1]. There are, therefore, several reported procedures for synthesis polysaccharide-based adsorbents.

4.1.1 Crosslinked Polysaccharides

The two most important factors controlling the extent of adsorption properties of a polysaccharide-based material are the hydrophilicity of the polymer and its crosslinking density [2]. Polysaccharides can be crosslinked by a reaction between the hydroxyl or amino groups in the chains with a coupling agent to form water-insoluble crosslinked networks [3–6]. Gels are formed when polysaccharides are crosslinked and these gels are separated into two classes of physical [7] and chemical gels [8]. Chemical gels are formed from irreversible covalent linkages, while physical gels are formed from various reversible linkages. Many researchers have carried out synthesis of crosslinked polysaccharides materials using different coupling agent.

Jiang et al. [9] prepared an Insoluble β-cyclodextrin (β-CD) using epichlorohydrin (EPI) as crosslinking agent under basic conditions. Earlier, Yamasaki et al. [10] prepared crosslinked insoluble β-CD polymer beads using hexamethylene diisocyanate (HDI) as the crosslinking agent. Some reports have suggested considerably more complex structures when EPI reacts with chitosan because the crosslinking agent could also react with amino groups of chitosan chains [11]. Recently, Mittal et al. [12] carried out graft copolymerization on Gum

© The Author(s) 2017
N.A. Oladoja et al., *Polysaccharides as a Green and Sustainable Resource for Water and Wastewater Treatment*, Biobased Polymers,
DOI 10.1007/978-3-319-56599-6_4

polysaccharides to produce hydrogels as adsorbents for wastewater purification. Nuclear Magnetic Resonance (NMR) relaxation study has shown that crosslinking of polysaccharides is not homogeneous and increasing the crosslinking degree in polysaccharide-based adsorbents increases the amorphous content of the product formed [13].

However, cross linked polymers can be obtained in homogeneous or heterogeneous conditions by using reticulation with bi- or poly-functional crosslinking agents such as epichlorohydrin [14–17], ethylene glycol diglycidyl ether [18–20], glutaraldehyde [21–24], benzoquinone [25], Maleic anhydride [26] or Isocyanates [2, 27]. All these crosslinking agents (with the exception of Maleic anhydride), are considered to be a hazardous environmental pollutant and potential carcinogens and cytotoxins. So they are neither safe for human health nor environmentally friendly, although they have the advantage that they do not eliminate the cationic amine function of a natural polymer like chitosan, which is the major adsorption site for pollutants during adsorption. Other water-soluble crosslinking agents have been proposed, such as sodium trimetaphosphate [30], sodium tripolyphosphate [28] calcium chloride [29], phosphorus oxychloride [30] or carboxylic acids [31]. Naturally, crosslinked materials possess several characteristic properties, and advantages such as:

- Ease of preparation i.e. homogeneous crosslinked materials which are easy to prepare with relatively inexpensive reagents and are available in a variety of structures with a variety of properties, and also in various configurations [32].
- Being insoluble in acidic and alkaline mediums as well as organic solvents.
- Retention of properties i.e. after crosslinking, they maintain their properties, fundamental characteristics (except the crystallinity) and strength in acidic and basic solutions. Furthermore, the swelling behaviour of the beads in aqueous solution can be optimized.
- Reduction of crystalline domains in the polysaccharide which can change the crystalline nature of the raw polymer. This parameter does influence significantly, the adsorption properties of the polymeric adsorbent because it does control accessibility to adsorption sites. Crosslinked materials also have other advantages such as faster kinetics, increased ease of operation and unusual diffusion properties [33]. Due to the hydrophilic nature of their crosslinking units, the adsorbent possesses a high ability to swell in water, and consequently, their networks are sufficiently expanded to allow a fast diffusion process for the uptake of the pollutants [32].
- Possibility of grafting functionalities on the polymer to suit desired application. Even though crosslinked polysaccharides have good adsorption properties, their adsorption capacities can be improved on by grafting various functional groups onto the polymer network or the polymer backbone [32–35].
- After adsorption, the crosslinked materials can also be easily regenerated by washing using a solvent or by solvent extraction as shown in Fig. 4.1.

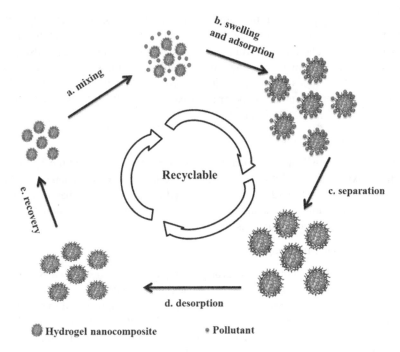

Fig. 4.1 Use of Hydrogels as Adsorbents in water treatment reproduced from Mittal et al. [36] Copyright © 2016 Wiley online Library

4.1.2   Polysaccharide Composites

Polysaccharide composites are formed from the immobilization of polysaccharides on insoluble supports by coupling or grafting reactions to give hybrid or composite materials. This method relies on the bonding of polysaccharides to a pre-existing mineral or organic matrix via several spacer arms. Composite materials may combine the physical and chemical properties of both inorganic and organic materials [3, 37–48]. Understandably, from literature, different kinds of substances have been used to form composites with polysaccharides, in particular with chitosan. These composites include montmorillonite [49], polyurethane [50], activated clay [51], bentonite [52], polyvinyl alcohol [53]. Chitosan composites have been proven to have better adsorption capacity and resistance to acidic environment [54]. The preparation is found to be simpler. Although Mittal et al. [36] have reviewed extensively on all types of polysaccharides, yet chitosan appears to be the most preferred polysaccharide by most researchers perhaps because it is easily sourced and in sufficient quantity. The various composites made from chitosan and how they are prepared are discussed below.

Chitosan/Clay composites

In preparing a chitosan/clay nanocomposite, the chitosan is dissolved by stirring in acetic acid. Dispersed clay is then added to the dissolved chitosan slowly in the appropriate ratio. The reaction mixture is then stirred for a period, centrifuged and washed. Wang et al. [49] found that the molar ratio of chitosan to montmorillonite could influence the chemical environment of the composite and hence its adsorption properties. The increment in the molar ratio of chitosan in a chitosan-montmorillonite composite increased its adsorption capacity of for Congo Red dye in aqueous solution until the molar ratio exceeds 1:1 after which the adsorption capacity remained almost constant.

Chitosan/Silica Composite

Composite chitosan-silica is obtained by the sol-gel method through the hydrolysis of tetraethoxysilane in chitosan solution. Acidified tetraethoxysilane is added dropwise to already prepared chitosan and stirred for 24 h using a magnetic stirrer. This mixture is allowed to stand for 7 days after which the solution becomes matured. The substance obtained is then dried at 60 °C [55].

Chitosan/activated clay composite

Chitosan/activated clay composite beads are prepared by mixing equal weights of chitosan flakes with clay activated in acetic acid and agitated for ca. 10 min. This mixture is dried in a vacuum oven for 3 h to remove air bubbles and then sprayed dried into a neutralization solution containing 15% NaOH and 95% ethanol in a volume ratio of 4:1. The composite beads formed are left standing for 1 d and afterwards washed with deionized water [51]. It is observed that a 1:1 ratio of activated clay to chitosan had a specific gravity of 1.0197 (g/cm^3) while that for the chitosan beads was 1.0055 (g/cm^3). Based on specific gravity values, Chang and Juang [51] concluded that the addition of activated clay could enhance the ability of chitosan to agglomerate and improve the hardness of the beads formed based on Stokes law. The hardness of the beads is necessary because it facilitates the separation of the adsorbents from solution without swelling. In the adsorption studies of Methylene blue dye and Reactive Red dye (RR222), chitosan composite beads had a comparable adsorption capacity with chitosan beads [52, 53]. However, the advantage of the beads is its ability to withstand attrition when used in a continuous flow mode to treat water.

Chitosan/polyurethane composite

Lee et al. [50] used polyurethane foam to prepare a composite with chitosan according to the reaction scheme shown in Fig. 4.2.

The preparation of chitosan/polyurethane composites is different from other methods. According to Lee et al. [50], at 0.25 wt% of glutaraldehyde concentration, chitosan was found to be better immobilized on the polyurethane matrix foam. With 20% chitosan content in the composite, it is observed that there is well-developed open-cell structures that enhances the accessibility of acid dyes into the composite adsorbent [50, 56]. A comparison between neat polyurethane foams and chitosan/polyurethane composites in the adsorption of acid dyes was made. It was observed that the adsorption capacity of neat polyurethane was comparably lower. It was

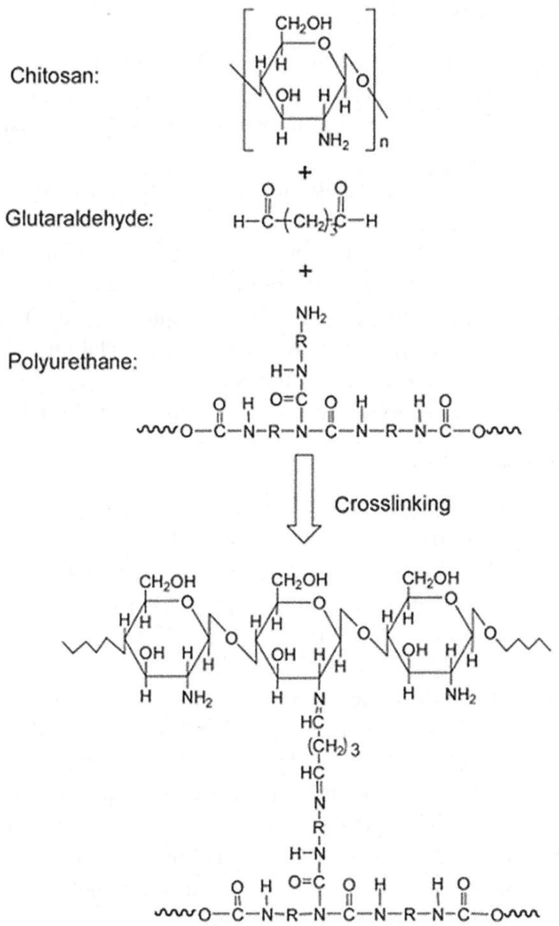

Fig. 4.2 Polyurethane/Chitosan composites prepared via crosslinking using glutaraldehyde. Reproduced from Lee et al. [50]

suggested that the amine groups in the neat polyurethane do not behave as an active site for the adsorption of the dyes [50].

Chitosan/cotton composite

In the preparation of the chitosan/cotton composite beads, cotton fiber is treated with sodium periodate ($NaIO_4$) before being added to the chitosan solution. Oxidation of carbohydrates by periodate ion (glycon cleavage) was for a long time, a classical method used for structure determination of complex carbohydrates. In later years, it seems more or less abandoned as more efficient methods were developed [57]. Periodates have been used to introduce dialdehydes to polysaccharides or glycoproteins [57]. Ethylene glycol is added to the solution after the cotton fiber is treated with sodium periodate to terminate the reaction [58].

Chitosan/sand composite

Chitosan/sand composite is prepared by mixing chitosan and sand (1:20% wt/wt) in 5% HCl for 5 h at room temperature. The resulting mixture is then neutralized with 1 M NaOH added to the mixture in a dropwise manner. Chitosan/sand composite precipitate is then washed and dried in a vacuum oven [15]. This composite has shown better adsorption capacity than any of its individual component used due to the three-dimensional structure of the composite adsorbent. Amine groups in chitosan provide active sites for the formation of complexes with metallic ions, which are stabilized by coordination. As the amount of acetyl groups in the structure increases, the maximum adsorption capacity for heavy metals removal from aqueous solution decreases [15]. In a recent work to improve the mechanical and chemical stability of chitosan/sand composite, chitosan was coated onto the sand (CCS) and cross-linked using Epichlorohydrin (ECH) and ethylene glycol diglycidyl ether (EGDE) [20].

Chitosan/cellulose composites

Cellulose is a polydisperse linear homopolymer consisting of regio-enantioselective β-1,4 glycosidic linked D-glucose units. The polymer contains three reactive hydroxyl groups at C-2, C-3 and C-6 atoms which are accessible to typical conversions [59]. Three hydroxyl groups in each β-D-glucopyranose units can interact with one another forming Intra- and intermolecular hydrogen bonds that gives rise to various types of supra-molecular semicrystalline structures [60]. The crystalline and also hydrogen bonding patterns have a strong influence on the chemical behaviour of cellulose. A consequence of this is the insolubility of macromolecule in water as well as in common organic liquids. This insolubility has stimulated the search for solvents appropriate for homogeneous phase reactions which still use alternative synthesis paths [59]. Modification of cellulose can be achieved through esterification [61, 62], halogenation [63], oxidation [64] and etherification [65]. However, there are some reports on the use of cellulose immobilized on chitosan to form chitosan–cellulose composite beads. Sun et al. [66] used an ionic liquid, 1-butyl-3-methylimidazolium, instead of acetic acid as the solvent for dissolution purpose. Ionic liquids (ILs) have recently received much attention as green solvents and are promising replacements for the traditional volatile organic solvents due to their characteristics such as non-volatility, nonflammability, thermal stability and ease of recycling [67, 68].

4.1.3   Nanoporous Polysaccharide Composites

The adsorption of different pollutants such as heavy metal ions and dyes from the contaminated water using nanocomposites has attracted significant attention due to their characteristic properties such as small size, large surface area, the absence of internal diffusion resistance, and high surface area-to-volume ratio. Mesoporous silica–cellulose hybrid composites were prepared by surface sol–gel coating process on natural cellulose substance [69]. The template, cetyltrimethylammonium bromide (CTAB) in the silica film can be removed by extraction to obtain high specific surface area (80.7 m^2 g^{-1}), which is two orders of magnitude greater than that of raw cellulose [69]. A novel micro/nanoporous superabsorbent hydrogel was synthesized by Soleyman et al. [70]. The authors developed a polysaccharide/protein based porous hydrogel by hydrolysis after the preparation of the hydrogels which involved free radical graft copolymerization of a combination of Salep, gelatin and acrylamide (AAm) in the presence of a crosslinking agent, N,N'- methylene bisacrylamide (MBA).

In the area of nanofibers, the conventional method of preparing nanoparticles from ultrafine polysaccharide nanofibers, such as cellulose nanofibers, includes the use of a strong acid or alkali pretreatment to remove pectin, hemicelluloses, and lignin, followed by a cryo-crushing defibrillation step. The usage of corrosive acids/alkali and special devices for defibrillation severely affects the practical value of this method for large-scale production [71]. However, a new method has been suggested, which involves the use of oxidation to treat polysaccharide pulps/powders with 2,2,6,6-tetramethylpiperidine-1-oxyl (TEMPO)/NaBr/NaClO in aqueous solution at ambient temperatures [72]. By way of making progress, this kind of ultrafine, (UF) polysaccharide nanofiber composite can be prepared from several sources including bleached wood pulp, microcrystalline R-cellulose from high-quality wood pulp, and chitin powders from crab shells, using the TEMPO/NaBr/NaClO method [73]. The unique properties of polysaccharide nanofibers, such as high crystallinity, high thermal stability, and high potential for functional modifications provide promise for the development of high flux UF membranes for energy-saving water purification process. Nanomaterials, in particular, nanocomposites, have a wide array of application including wastewater treatment.

4.2 Treatment of Water and Wastewater

4.2.1 Pollutant Removal

Water is major natural resource of planet earth; the volume of its consumption in the world amounts to billions of cubic meters per year [74]. Therefore, there is the need to consider its use and its protection including research directed towards working out innovative methods and materials for water quality preservation [75].

Water pollution due to toxic inorganic and organic compounds remains a serious environmental and public challenge. Moreover, with more and more stringent regulations, water pollution has also become a major source of concern and a priority for most industrial sectors. Heavy metal ions, aromatic compounds (including phenolic derivatives, and polycyclic aromatic compounds) and dyes are present in the environment as a result of their extensive industrial uses. They are common contaminants in wastewater, and many of them are known to be toxic or carcinogenic. For example, chromium (VI) is found to be toxic to bacteria, plants, animals and humans [76]. Mercury and cadmium are known as two of the most toxic metals that are very damaging to the environment [77]. Also, heavy metals are not biodegradable and tend to accumulate in living organisms, causing various diseases and disorders. Therefore, their presence in the environment, in particular in water, should be controlled. Chlorinated phenols are also considered as priority pollutants since they are harmful to organisms even at low concentrations. They have been classified as hazardous pollutants because of their high malignant potential with respect to human health [78]. For example, 2,4,6-trinitrotoluene (TNT) is a nitro-aromatic molecule that has been widely used by the weapons industry for the production of bombs and grenades. This compound is recalcitrant, toxic and mutagenic to various organisms [79, 80]. Many synthetic dyes, which are extensively used for textile dyeing, paper printing and as additives in petroleum products are recalcitrant organic molecules that strongly colour waste water.

Strict legislation on the discharge of these toxic products makes it necessary to develop various efficient technologies for the removal of pollutants from wastewater. Different techniques and processes are currently used for the removal of pollutants from wastewater. Some of these techniques include biological treatments [81–83], membrane processes [84, 85] advanced oxidation processes [86–89], electrochemical advanced oxidation processes [90], and adsorption procedures [91–94]. However, adsorption technique is the most widely used for removing inorganic and organic compounds from water and wastewater.

4.2.1.1   Inorganic Pollutants

Inorganic pollutants, salts and metals, can be naturally-occurring or can be from urban stormwater runoff, industrial or domestic wastewater discharges, oil and gas production, mining, or farming. Most inorganic chemicals polluting aquatic environments are heavy metals, inorganic anions and radioactive materials. This pollution could be caused by naturally occurring substances such as fluoride, arsenic and boron; by industrial waste containing mercury, cadmium, chromium, cyanide and others; by agricultural and domestic waste containing nitrogen compounds, and from contamination by copper, iron, lead and zinc during the distribution of drinking water using lead and zinc pipes. Naturally occurring inorganic substances mainly contaminate groundwater, whereas industrial and agricultural waste contaminates surface water such as rivers, lakes and ponds. The primary source of contamination in drinking water is the distribution system [95].

To mitigate the levels of these pollutants in water several polysaccharide composites have been used. Chitosan/cotton composites has been used to adsorb Hg^{2+} [96], Pb^{2+}, Ni^{2+}, Cd^{2+}, Cu^{2+} [97] and Au^{3+} [98]. Chitosan/Montmorillonite was prepared to remove As^{3+} [99] and Cu^{2+} [100]. This chitosan/sand composite crosslinked with CCS-EGDE was used to adsorb Cu^{2+} [20].

Glutaraldehyde-crosslinked chitosan beads was used for the adsorptive separation of Au^{3+} and Pd^{2+} [101]; cross-linked chitosan-poly (aspartic acid) chelating resin containing disulfide bond for the adsorption of Pb^{2+} and Hg^{2+} [102]; and magnetic carboxymethylchitosan nanoparticles for adsorption of Pb^{2+}, Cu^{2+}, Zn^{2+} [103], chitosan modified zeolite for the removal of ammonium and phosphate [104] to mention but a few.

4.2.1.2 Organic Pollutants

Organic pollution is the term used when large quantities of organic compounds are present in a system. It originates from domestic sewage, urban run-off, industrial effluents and agriculture wastewater, sewage treatment plants and industries including food processing, pulp and paper making, agriculture and aquaculture industries. During the decomposition process of organic pollutants, the dissolved oxygen in the receiving water may be consumed at a greater rate than it can be replenished, causing oxygen depletion and having severe consequences for the stream biota. Wastewater with organic pollutants contains large quantities of suspended solids which reduce the light available to photosynthetic organisms and, on settling out, alter the characteristics of the river bed, rendering it an unsuitable habitat for many invertebrates.

Organic pollutants also include pesticides, fertilizers, hydrocarbons, phenols, plasticizers, biphenyls, detergents, oils, greases, pharmaceuticals, proteins and carbohydrates [105, 106]. Toxic organic pollutants cause several environmental problems and health issues. The most common organic pollutants are the persistent organic pollutants (POPs). POPs are compounds of high concern due to their toxicity, persistence, long-range transportability [107] and bioaccumulation in animals [108]. POPs are carbon-based chemical compounds and a mixture of twelve pollutants that include industrial chemicals such as polychlorinated biphenyls (PCBs), polychlorinated dibenzo-p-dioxins and dibenzofurans (PCDD/Fs), and some organochlorine pesticides (OCPs), such as hexachlorobenzene (HCB) or dichloro-diphenyl-trichloroethane (DDT), dibenzo-p-dioxins (dioxins) and dibenzo-p-furans (furans) [109]. PCDD/Fs are released into the environment as by-products of several processes, like waste incineration or metal production [110]. Many of these compounds are still being used in large quantities especially in the developing countries. Due to their environmental persistence, they have the ability to bioaccumulate and biomagnify [111] leading to health issues in man.

Polysaccharide-based adsorbents have been used to remove organics from aqueous solution mainly dyes. Chitosan-glutaraldehyde copolymer was used to adsorb phenol from solution [112], organo-arsenical [113], and para-nitrophenol [114]. Magnetic chitosan was used for the adsorption of Diclofenac and Chlofibric acid [58]; β-cyclodextrin/cross-linked chitosan for adsorption of phenols [115]; chitosan grafted with sulfuric acid for adsorption of Pramipexole [116]; Graphite oxide/Carboxyl-grafted chitosan for Dorzolamide [117]; amphoteric cellulose for adsorption of anionic dyes [118]; epichlorohydrin cross-linked maize- and corn-derived starches for the adsorption of para-nitrophenol [47]; and magnetic chitosan functionalized with graphite oxide for adsorption of reactive black 5 dye [46]. Figure 4.3 shows the mechanism of synthesis of graphite oxide/magnetic chitosan composite (GO-Chm) after functionalization of magnetic chitosan (Chm) with graphite oxide (GO). The proposed interactions of the Reactive Black 5 (RB5) with the prepared GO-Chm is also shown in Fig. 4.3.

4.2.2 Adsorption Mechanism

The deposition of a particular component on the surface or at the interface between two phases is known as adsorption. Adsorption is a surface phenomenon. In this process, there are two components one of which is the *adsorbent*-the solid surface and the compound (pollutant) that sticks or gets attached to the solid surface which is called the *adsorbate*. The adsorption capacity of the adsorbent may be altered with change in solution temperature, nature and concentration of pollutants, presence of other pollutants and other experimental conditions such as pH, contact time and particle size of the adsorbent. The presence of suspended particles, oil and grease also affect the efficiency of the process. Therefore, prefiltration is sometimes required. When an adsorbent interacts with the polluted water, the pollutants adhere to the surface of the adsorbent and equilibrium is established. The first major challenge in adsorption research is to select the most promising type of adsorbent from a vast pool of readily available materials. The next real challenge is to identify the adsorption mechanism(s). In general, there are three steps required for pollutant adsorption onto the solid adsorbent:

   (i)   The transport of the pollutant from the bulk solution to the adsorbent surface;
   (ii) Adsorption on the particle surface; and
   (iii) Transport within the adsorbent particle.

Adsorption studies, in particular, kinetics and isotherm studies provide information on the mechanism of adsorption which attempts to explain how the pollutant is bound onto or within the adsorbent. This knowledge is essential for understanding the adsorption process and for selecting the desorption strategy. Due to the complexity of materials used and their specific characteristics (such as the presence of complexing chemical groups, small surface area, and poor porosity), the adsorption mechanism of polysaccharide-based adsorbents is different from those of other conventional adsorbents. These mechanisms are, in general, complicated because they sometimes involve several chemical and physical interactions simultaneously [41, 119]. Also, a wide range of chemical structures, pH, salt concentrations and the presence of ligands often add to the complication. Some of the reported interactions include Ion-exchange, complexation, coordination/chelation, electrostatic interactions, acid–base interactions, hydrogen bonding, hydrophobic interactions, physical adsorption and precipitation.

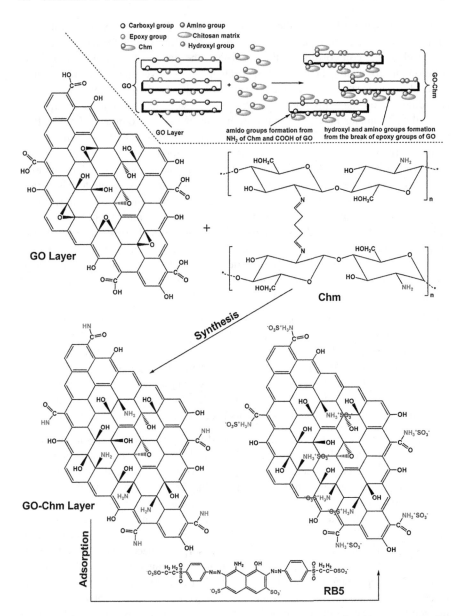

Fig. 4.3 Proposed mechanism of synthesis of graphite oxide/magnetic chitosan composite (GO-Chm) after functionalization of magnetic chitosan (Chm) onto graphite oxide (GO). Proposed interactions of the Reactive Black 5 (RB5) adsorption onto the prepared GO-Chm. Reproduced from Travlou et al. [46], Copyright © 2013 American Chemical Society

An examination of data in the literature indicates that it is possible that at least some of these mechanisms are to varying degrees acting simultaneously depending on the chemical composition of the adsorbent, the nature of the pollutant and the solution environment. The removal of metal species or dyes from aqueous solution using chitosan is strongly dependent on pH [22, 119, 120] which may thus involve two different mechanisms (chelation versus ion exchange) depending on the pH since this parameter may affect the protonation of the macromolecule [120]. It is well known that chitosan is characterized by its high percentage of nitrogen, present in the form of amine groups that are responsible for metal ion binding through chelation mechanisms. Amine sites are the main reactive groups for metal ions although hydroxyl groups, especially at C-3 position, may contribute to adsorption [119, 120]. However, chitosan is also a cationic polymer, and its pKa ranges from 6.2 to 7.0 depending on the deacetylation degree and the ionization extent of the polymer [34]. Thus, in acidic solutions, it is protonated and possesses electrostatic properties. It is therefore also possible to adsorb metal ions through anion exchange mechanisms [22, 119]. Although several contradictory mechanisms have been proposed, metal adsorption on chitin and chitosan derivatives is now assumed to occur through several single or mixed interactions including

(i)   chelation interaction (coordination) on amino groups in a pendant fashion or in combination with vicinal hydroxyl groups,
(ii)   complexation phenomena (electrostatic attraction) in acidic media,
(iii) ion-exchange with protonated amino groups through proton exchange or anion exchange, the counter ion being exchanged with the metal anion [34, 119]. Physical adsorption plays little role in the interaction between cross-linked chitosan beads and pollutants because beads have a small surface area.

The pH may also affect the speciation of metal ions, changing the speciation of the metal resulting in a change from chelation mechanism to electrostatic attraction mechanism. Another parameter that can play a major role in the mechanism is the presence of ligands grafted on the chitosan chains. For crosslinked starch materials, physical adsorption in the polymer structure and chemisorption of the pollutant via hydrogen bonding, acid–base interactions, complexation, and ion exchange are both involved in the adsorption process [16, 121]. In most cases, though a combination of these interactions was proposed to explain adsorption mechanisms, the efficiency and the selectivity of these adsorbents are mainly attributed to their chemical network. When the materials, for example, contain cyclodextrin (CD) molecules, the mechanism is due to the formation of an inclusion complex between the CD molecule and the pollutant through host–guest interactions. It has also been reported that the presence of pollutant–pollutant hydrophobic interactions can also be used to explain the adsorption process [16, 122].

Adsorption mechanisms with polysaccharide-based adsorbents are even more complex, since not only does the polysaccharide play a role in the adsorption but also the other components of the adsorbents. However, recent studies have indicated that, in general, a given pollutant-binding mechanism may be explained by its interaction with a component of the composite matrix. For example, in the grafted or coated silica-based sorbents containing biopolymers, the formation of interactions between the polymer and the pollutant mainly explain the mechanism although the mineral matrix may contribute to the adsorption [123, 124]. Crini et al. [37, 125] demonstrated that in the case of organic beads containing amine functional groups grafted by cyclodextrins used as adsorbents, the adsorption mechanism is due to the formation of an inclusion complex between the cyclodextrin molecules and the pollutant through host–guest interactions. However, even though the cyclodextrins play a dominant role, physical adsorption in the resin polymer network and chemical interactions via acid–base interactions, ion exchange and hydrogen bonding due to the amino groups are also involved in the adsorption process. These interactions are significant. The authors also showed that the adsorption is strongly dependent on pH. The nature of the bonds between pollutants and these adsorbents' surfaces depends on the extent of the acid–base interactions. For weak acid–base interactions, only hydrogen bonds may be formed. For strong acid–base interactions, the interactions may gradually change to a chemical complexation. The presence of co-existing interactions between the pollutants and the sorbent matrix should be taken into account for the regeneration of the loaded adsorbent.With chitosan modified zeolite it is suggested that the negative charge in the internal pores of the zeolite, the oxides of CaO, Al_2O_3 and Fe_2O_3 in the non-zeolite fraction, and the monolayer chitosan on the external surface of the zeolite are involved in the retention of cationic ammonium, anionic phosphate, and organic humic acid, respectively on its surface [104].

4.2.3 Regeneration Techniques

The regeneration of the sorbent is vital in keeping the process costs down and to open up the possibility of recovering the pollutant extracted from the solution. For this purpose, it is desirable to desorb the adsorbed pollutants and to regenerate the material for use in several cycles of application. The desorption of the process is expected to yield the pollutants in their concentrated form and restore the adsorbent close to its original condition for efficient reuse with undiminished pollutant uptake efficiency and no physical change or damage to the adsorbent material. Given the developments of the various kinds of adsorbents described in literature, the stability of the materials under adsorption process conditions and the reproducibility of the adsorption properties of the adsorbent is of utmost importance.

There are reviews on regenerating inorganic and organic pollutant loaded on various types of adsorbents [126, 127]. Unfortunately, there is the dearth of information on the regeneration of spent polysaccharide-based adsorbents, their stability, and reproducibility. However, there are a few report on the regeneration of polysaccharide-based adsorbents by some researchers. Crini [122] demonstrated the reproducibility of the adsorption properties of cyclodextrin polymers used as adsorbents for the removal of various dyes and the regeneration of the adsorbents after saturation. Since the interactions between the pollutant and cyclodextrin are driven mainly by hydrophobic interactions, organic solvents are good candidates for the regeneration of the material. As a result, the polymer adsorbents were easily regenerated using ethanol as the washing solvent. The adsorption capacity value remained unchanged after this treatment. This unchanged adsorption capacity was as a result of the chemical stability of the crosslinked gels. Furthermore, Delval et al. [16], showed that crosslinked starches loaded with dye molecules, were easily regenerated using Soxhlet extraction with ethanol rather than with water, and the adsorption capacity of the adsorbent in batch experiments remained constant. Kim and Lim [26], reported that crosslinked-starch derivatives could be easily regenerated by washing with a weak acid solution.

Modified Magnetite nanoparticles were used and regenerated over five cycles when loaded with Cu^{2+}, Zn^{2+}, and Pb^{2+} and its adsorption capacity was not altered significantly [128]. Crosslinked chitosan beads and Magnetic Carboxymethylchitosan Nanoparticles were easily regenerated using EDTA (a potent chelating agent) solutions [129]. The free electron doublet of nitrogen on amine groups is known to be responsible for the adsorption of metal cations on chitosan derivatives and the adsorption usually, occurs at a pH close to neutral. The complexation of the pollutant by the EDTA ligands displaces the pollutant from the material.

Regeneration could also be effected by contact with mildly strong acidic solution. This change in the pH of the solution reverses the adsorption because adsorption by chelation mechanism is very sensible to pH as shown by Onditi et al. [130] who used a 0.1 M HCl to desorb Pb^{2+} and Cd^{2+} from a polysaccharide extract isolated from cactus pads. Cyclodextrin modified silica beads were reported to keep a constant capacity towards the removal of phenolic compounds after several cycles of adsorption and regeneration with methanol [123, 124].

However, the question of the long-term stability of these polysaccharide-based adsorbents especially when exposed to various chemical environments, has raised concerns for some time now depending on whether these adsorbents are prepared by grafting or coating. In the former case, the main concern is the stability of the covalent bond between the polysaccharide and the matrix. In spite of rather good results regarding adsorption capacity, some stability problems were encountered with polysaccharide grafted silicas especially when regenerated [124]. However, with the coating method, polysaccharide-based adsorbent stability depends on the strength of the interactions between the biopolymer and the silica surface. To increase the stability and to avoid desorption of the biopolymer from the matrix, several researchers have proposed a crosslinking reaction, after the coating of the polymer. Several crosslinking agents are available including 4,4-methylene diphenyl diisocyanate (MDI), hexamethylene 1,6 diisocyanate (HDI) and epichlorohydrin (EPI) [129, 130]. Mallard et al. [131] has proposed that in the the functionalization of zeolite Y (Faujasite) with β-cyclodextrin (CD) for the removal of toluene and methyl orange dye from aqueous solutions, a more efficient adsorbent is produced when an intial grafting of the linker (3-glycidoxypropyltrimethoxysilane) is done on the β-CD and then reacted with zeolite than grafting of the linker on zeolite and then reacting with β-CD.

4.2.4 Polysaccharide-based Adsorbents Versus Other Adsorbents

Generally, a suitable adsorbent for adsorption processes of pollutants should meet several requirements which inlcudes—(i) efficient for removal of a wide variety of target (hydrophobic) pollutants (ii) high capacity and rate of adsorption (iii) significant selectivity for adsorbate at different initial concentrations (iv) granular type with good surface area (v) high physical strength (vi) able to be regenerated if required (vii) tolerant for a broad range of wastewater parameters (viii) and low cost.

Polysaccharide-based adsorbents for adsorption of pollutants from solution, compared with other adsorbents, such as commercial activated carbons and synthetic ion-exchange resins, offers several advantages:

1. Polysaccharide-based adsorbents are low-cost materials obtained from raw natural resources. The majority of commercial polymers and ion-exchange resins are derived from petroleum-based raw materials using chemical processes that are not always safe and environmentally friendly.

2. The use of polysaccharide-based adsorbents is extremely cost-effective. Crosslinked adsorbents are easy to prepare with relatively inexpensive reagents (low operating cost). The amount of adsorbent used is, in general, reduced as compared to conventional adsorbents since they are more efficient. Much smaller quantities of biopolymer are needed to build the hybrid adsorbents. However, activated carbon and synthetic ion exchange resins are quite expensive, and the higher the quality, the greater the cost. Besides, crosslinking does impact some level of strength on polysacchride-based adsorbent, reducing its attrition ability.

3. Polysaccharide-based adsorbents are versatile. This versatility allows the adsorbent their use under different forms, from insoluble beads to gels, sponges, capsules, films, membranes or fibres which is not so with other types of adsorbents. Polysaccharide-based adsorbents are available in a variety of structures with a variety of properties. The utilization of cross-linked beads or hybrid materials has many advantages regarding applicability in a broad range of process configurations and reusability for repeated cycles of adsorption-desorption [132]. Tuning the properties of the polysaccharide-based adsorbent to meet certain application demands is very possible unlike with activated carbon and resins.

4. Adsorbents based on polysaccharides are very efficient for the removal of pollutants at different concentrations. They possess a high capacity and high rate of adsorption, high efficiency, and selectivity in detoxifying both very dilute and concentrated solutions. In general, activated carbon and synthetic resins suffer from a lack of selectivity and applications are typically limited to levels of contaminants in the parts per million (ppm) range. Biosorption on polysaccharide derivatives, in particular, chitosan, is an emerging technology that

attempts to overcome the selectivity disadvantage of the adsorption processes. Chitosan beads are often more selective than traditional resins and can reduce pollutant concentrations to ppb levels.

5. With repetitive functional groups, biopolymers provide excellent chelating and complexing materials for a wide variety of pollutants including dyes, heavy metals, and aromatic compounds. Moreover, even though polysaccharides and their derivatives have high adsorption capacity in their natural state, their adsorption capacity can be improved only by the substitution of various functional groups onto the polymer backbone.

6. Cross-linked polysaccharide-based adsorbents, in particular, cyclodextrin polymers, possess an amphiphilic character. It is precisely this nature of these adsorbents that makes them so appealing since they are hydrophilic enough to swell considerably in water allowing fast diffusion processes for the adsorbates, while they possess highly hydrophobic sites, which efficiently trap non-polar pollutants. Synthetic resins have a weak contact with aqueous solutions and their modification and pretreatment by activation solvents are necessary for enhanced water wettability. Activated carbons adsorb poorly some some non-polar pollutants.

7. The regeneration step for pollutant-loaded polysaccharide-based adsorbents is easy. The adsorbents can be regenerated by desorption at low cost if required. They are easily regenerated by a washing solvent since the interaction between the pollutant and sorbent is driven mainly by electrostatic, hydrophobic and ion-exchange interactions. The desorption process produces the pollutant in a concentrated form and restores the material close to the original condition for efficient reuse with no physical–chemical changes or damage. The regeneration of saturated carbon by the thermal and chemical procedure is known to be expensive, and results in loss of the adsorbent and thus adsorption capacity of the adsorbent via attrition.

Despite the number of papers published on natural adsorbents for pollutant uptake from contaminated water, there is still very little literature containing a full study of comparisons between adsorbents. In fact, the data obtained from the biopolymer derivatives have not been compared systematically with commercial activated carbons or synthetic ion-exchange resins, which showed high removal efficiencies, except in recent publications. The results reported by Chiou et al. [17] demonstrated that the adsorption capacities of cross-linked chitosan derivatives toward anionic dyes are much higher (3–15 times) than commercial activated carbons at the same pH. The beads exhibit excellent performance and appear to be much more efficient and selective. Coughlin et al. [133] concluded that adsorption on chitosan is competitive against precipitation techniques. Ngah and Isa [134] found that adsorption capacities were significantly greater for chitosan in the recovery of copper ions than commercial synthetic resin. Martel et al. [135] demonstrated that the chitosan beads containing cyclodextrin are characterized by both a rate of sorption and efficiency superior to that of the parent chitosan bead without CD and crosslinking cyclodextrin–epichlorohydrin gels.

However, it must be said here that the uptake capacity of two adsorbents for the removal of pollutants from aqueous solution should be compared only at the same equilibrium concentration of the adsorbate and/or similar reaction/process conditions. The comparison of adsorption performance also depends on several other parameters related to the effluent (competition between pollutants) and the analytical methods (batch method, column, dynamics tests, reactors) used for decontamination tests [32, 136–139]. Thus, a direct comparison of data obtained using different polysaccharide-based materials (as is common today) is not appropriate since experimental conditions for their application in adsorption of pollutants, existing in literature are not entirely the same. Also, due to the scarcity of consistent cost information, price comparisons for the different polysaccharide-based adsorbents are also difficult to make.

4.3 Prospects and Challenges

There are, of course, challenges in using polysaccharide-based adsorbents in waste water treatment. To identify these challenges will be useful for future research in this area and will further hasten the utilization of these polysaccharide-based adsorbents on a large scale. Some of the prospects and challenges to be encountered with using polysaccharide-based adsorbents are summarized as follows:

1. The adsorption properties of an adsorbent depend on the different sources of raw materials. The adsorption capacity of biopolymers like chitin and chitosan materials depends on the origin of the polysaccharide, molecular weight and solution properties, crystallinity, affinity for water, percent deacetylation and amino group content [140]. These parameters, determined by the conditions selected during the preparation [8], control the swelling and diffusion properties of the polysaccharide and have a strong influence on its characteristics. These problems can explain why it is difficult to apply polysaccharide-based adsorbents on an industrial-scale. It is also well-known that cyclodextrin possesses well-defined geometry, and the adsorption capacity of this biopolymer will be influenced not only by shape, size, and polarity of pollutants but also by the size of the cyclodextrin cavity.

2. The extreme variability of industrial wastewater should be taken into account in the design of any polysaccharide-based adsorbents. Each type of pollutant may require specific polysaccharide-based adsorbents. Similarly, each polysaccharide has its specific application as well as inherent advantages and disadvantages in wastewater treatment. Chitosan-based adsorbents, for example, have high affinities for heavy metal ions but low affinity for anionic dyes while cyclodextrin has a remarkable capacity to form inclusion complexes with organic molecules, especially aromatics, but low affinity for the metal ion. Therefore, the choice of the adsorbent depends on the nature of the pollutant. However, there is the possibility of developing a bifunctional

polysaccharide-based adsorbents in the near future (with negative and positive charges) that could adsorb, from a cocktail solution, both cationic and anionic pollutants.

3. The performance of a polysaccharide-based adsorbent is dependent on the type of material used, the extent of chemical activation and surface engineering or modifications carried out. Based on the nature of the substituent on the adsorbent and the degree of substitution, the properties of the adsorbent can be varied extensively. For grafted polymer materials, pollutant adsorption was found to be dependent on the degree of attachment of the polyfunctional groups. In the case of chitosan, the best method of achieving selective extraction is to use a metal specific ligand. However, it has proved impossible to find specific ligands for each metal ion. This remains a challenge for future research.

4. The efficiency of adsorption depends on physicochemical characteristics such as porosity, specific surface area and particle size of the polysaccharide-based adsorbent. Though these could be improved on if the polysaccharide-based adsorbent is properly engineered. For example, chitosan has a very low specific surface area between 2 and 30 $m^2 g^{-1}$. Glutaraldehyde cross-linked chitosan beads [25], EPI–cyclodextrin gels [45] and EPI–starch beads [16] have specific surface areas of ca. 60, 213 and 350 $m^2 g^{-1}$ respectively. However, most commercial activated carbons have a specific area of the order of 800–1500 m/g. Adsorption capacity increases with a decrease in the size of the particle since the effective surface area is higher for the same mass of smaller particles, and the time required to reach the equilibrium in an adsorption reaction is significantly increased with increase in the size of adsorbent particles. To improve on the surface area of polysaccharide-based adsorbents, there is need to make them highly porous and thus selective for purification purposes.

5. Pollutant molecules have many different and complicated structures which is one of the most important factors influencing adsorption. There is yet little information in the literature on this topic. Further research is needed to establish the relationships between pollutant structure and adsorbent at the molecular level so as to determine the precise mechanism of adsorption of these pollutants by polysaccharide-based adsorbents. This knowledge is essential for selecting the desorption strategy.

6. The production of chitosan involves a chemical deacetylation process. Commercial production of chitosan by deacetylation of crustacean chitin with strong alkali appears to have limited potential for industrial acceptance because of a significant amount of concentrated alkaline solution that is produced as waste in the process which is not environmentally friendly. There is, therefore, an urgent need to seek other sustainable and eco-friendly pathways of producing chitosan which is one of the most important polysaccharides used for preparing polysaccharide-based adsorbents.

7. With the increasing presence of pathogens in potable water, it will be valuable to intensify efforts in improving the disinfection properties of polysaccharide-based adsorbents as a sustainable but low-cost means of treating water. This will reduce the mortality rate of children under five years of age in developing

countries who are the most vulnerable to diseases resulting from ingestion of pathogenic polluted water. The insertion of cheaper and less toxic biocidal metals like Zn and Cu into the structure of this class of adsorbents can improve its disinfection properties rather than the current use of more expensive and more toxic metal like Ag. However, research should be carried out to investigate any eventual release of these metals into treated water at even the parts per billion (ppb) levels.

4.4 Conclusion

The use of polysaccharide-based adsorbents for water treatment is promising but requires further research. With the abundance of some of these polysaccharides in nature and their high adsorption capacities for the removal of pollutants from aqueous solution in addition to their rapid kinetics with respect to uptake of pollutants from aqueous solution, it will be interesting to see these adsorbents being utilized on a large scale in the form of highly porous and/or highly charged particles. Making them into membranes with less fouling ability will provide an alternative path to the development of highly efficient membranes for water treatment in the next decade. Research efforts in this direction will feed into the current United Nations Sustainability agenda.

References

1. R. S. Blackburn, Natural polysaccharides and their interactions with dye molecules: Applications in effluent treatment. Environ. Sci. Technol. **38**, 4905–4909 (2004)
2. O. Guven, M. Sen, E. Karadag, D. Saraydin, A review on the radiation synthesis of copolymeric hydrogels for adsorption and separation purposes. Radiat. Phys. Chem. **56**, 381–386 (1999)
3. G. Crini, M. Morcellet, Synthesis and applications of adsorbents containing cyclodextrins. J. Sep. Sci. **25**, 789–813 (2002)
4. L. Janus, B. Carbonnier, A. Deratani, M. Bacquet, G. Crini, J. Laureyns, New HPLC stationary phases based on (methacryloyloxypropyl-β-cyclodextrin-co-N-vinylpyrrolidone) o-polymers coated on silica. Preparation and characterisation. New J. Chem. **27**, 307–312 (2003)
5. A.C. Chao, S.S. Shyu, Y.C. Lin, F.L. Mi, Enzymatic grafting of carboxyl groups onto chitosan—to confer on chitosan the property of a cationic dye adsorbent. Bioresour. Technol. **91**, 157–162 (2004)
6. A.J. Varma, S.V. Deshpande, J.F. Kennedy, Metal complexation by chitosan and its derivatives: a review. Carbohydr. Polym. **55**, 77–93 (2004)
7. J. Berger, M. Reist, J.M. Mayer, O. Felt, N.A. Peppas, R. Gurny, Structure and interactions in covalently and ionically crosslinked chitosan hydrogels for biomedicals applications. Eur. J. Pharm. Biopharm. **57**, 19–34 (2004)

8. J. Berger, M. Reist, J.M. Mayer, O. Felt, N.A. Peppas, R. Gurny, Structure and interactions in chitosan hydrogels formed by complexation or aggregation for biomedical applications. Eur. J. Pharm. Biopharm. **57**, 35–52 (2004)

9. H. Jiang, Z. Yang, X. Zhou, Y. Fang, H. Ji, Immobilization of β-cyclodextrin as insoluble β-cyclodextrin polymer and its catalytic performance. Chin. J. Chem. Eng. **20**(4), 784–792 (2012)

10. H. Yamasaki, Y. Makihata, K. Fukunaga, Preparation of crosslinked β-cyclodextrin polymer beads and their application as a sorbent for removal of phenol from wastewater. J. Chem. Technol. Biotechnol. **83**, 991–997 (2006)

11. G. Crini, Recent developments in polysaccharide-based materials used as adsorbents in wastewater treatment. Progress Polym. Sci. **30**, 38–70 (2005)

12. H. Mittal, R. Jindal, B.S. Kaith, A. Maity, S.S. Ray, Synthesis and flocculation properties of gum ghatti and poly (acrylamide-co-acrylonitrile) based biodegradable hydrogels. Carbohydr. Polym. **114**, 321–329 (2014)

13. D. Shiftan, F. Ravenelle, M.A. Mateescu, R.H. Marchessault, Change in the V/B polymorph ratio and T relaxation of epichlorohydrin crosslinked high amylose starch excipient. Starch/Staerke **52**, 186–195 (2000)

14. M.S. Chiou, H.Y. Li, Equilibrium and kinetic modeling of adsorption of reactive dye on cross-linked chitosan beads. J. Hazard. Mater. **B93**, 233–248 (2002)

15. M.W. Wan, C.C. Kan, C.H. Lin, D.R. Buenda, C.H. Wu, Adsorption of copper (II) by chitosan immobilized on sand. Chia-Nan Annu. Bull. **33**, 96–106 (2007)

16. F. Delval, G. Crini, J. Vebrel, M. Knorr, G. Sauvin, E. Conte, Starch-modified filters used for the removal of dyes from waste water. Macromol. Symp. **203**, 165–171 (2003)

17. M.S. Chiou, P.Y. Ho, H.Y. Li, Adsorption of anionic dyes in acid solutions using chemically cross-linked chitosan beads. Dyes Pigm. **60**, 69–84 (2004)

18. X. Zeng, E. Ruckenstein, Cross-linked macroporous chitosan anion-exchange membranes for protein separations. J. Membr. Sci. **148**, 195–205 (1998)

19. F.L. Mi, S.S. Shyu, C.T. Chen, J.Y. Lai, Adsorption of indomethacin onto chemically modified chitosan beads. Polymer **43**, 757–765 (2002)

20. K.-J. Hsien, C.M. Futalan, W.-C. Tsai, C.-C. Kan, C.-S. Kung, Y.-H. Shen, M.-W. Wan, Adsorption characteristics of copper(II) onto non-crosslinked and cross-linked chitosan immobilized on sand. Desalin. Water Treat. **51**(28–30), 5574–5582 (2013)

21. E. Guibal, A. Larkin, T. Vincent, J.M. Tobin, Chitosan sorbents for platinum sorption from dilute solutions. Ind. Eng. Chem. Res. **38**, 4011–4022 (1999)

22. M. Ruiz, A.M. Sastre, E. Guibal, Palladium sorption on glutaraldehyde-crosslinked chitosan. React. Funct. Polym. **45**, 155–173 (2000)

23. M.S. Dzul-Erosa, M.T.I. Saucedo, M.R. Navarro, R.M. Avila, E. Guibal, Cadmium sorption on chitosan sorbents: kinetic and equilibrium studies. Hydrometallurgy **61**, 157–167 (2001)

24. R.S. Juang, H.J. Shao, A simplified equilibrium model for sorption of heavy metal ions from aqueous solutions on chitosan. Water Res. **36**, 2999–3008 (2002)

25. B.J. McAfee, W.D. Gould, J.C. Nadeau, A.C.A. Da Costa, Biosorption of metal ions using chitosan, chitin, and biomass of Rhizopus oryzae. Sep. Sci. Technol. **36**, 3207–3222 (2001)

26. T. Girek, D.H. Shin, S.T. Lim, Polymerization of β-cyclodextrin with maleic anhydride and structural characterization of the polymers. Carbohydr. Polym. **42**, 59–63 (2000)

27. K.P. Lee, S.H. Choi, E.N. Ryu, J.J. Ryoo, J.H. Park, Y. Kim, Preparation and characterization of cyclodextrin polymer and its high-performance liquid-chromatography stationary phase. Anal. Sci. **18**, 31–34 (2002)

28. H.J. Chung, K.S. Woo, S.T. Lim, Glass transition and enthalpy relaxation of cross-linked corn starches. Carbohydr. Polym. **55**, 9–15 (2004)

29. R. Dubey, J. Bajpai, A.K. Bajpai, Chitosan-alginate nanoparticles (CANPs) as potential nanosorbent for removal of Hg (II) ions. Environ. Nanotechnol. Monit. Manage. **6**, 32–44 (2016)

30. B.S. Kim, S.T. Lim, Removal of heavy metal ions from water by cross-linked carboxymethyl corn starch. Carbohydr. Polym. **39**, 217–223 (1999)

31. C. Seidel, W.M. Kulicke, C. Heb, B. Hartmann, M.D. Lechner, W. Lazik, Influence of the cross-linking agent on the gel structure of starch derivatives. Starch/Staeke **53**, 305–310 (2001)

32. A. Shweta, P. Sonia, Pharamaceutical relevance of crosslinked chitosan in microparticulate drug delivery. Int. Res. J. Pharm. **4**, 45–51 (2013)

33. S.E. Bailey, T.J. Olin, R.M. Bricka, D.D. Adrian, A review of potentially low-cost sorbents for heavy metals. Water Res. **33**, 2469–2479 (1999)

34. M.L. Arrascue, H.M. Garcia, O. Horna, E. Guibal, Gold sorption on chitosan derivatives. Hydrometallurgy **71**, 191–200 (2003)

35. C. Jeon, W.H. Höll, Chemical modification of chitosan and equilibrium study for mercury ion removal. Water Res. **37**, 4770–4780 (2003)

36. H. Mittal, S.S. Ray, M. Okamoto, Recent progress on the design and applications of polysaccharide-based graft copolymer hydrogels as adsorbents for wastewater purification. Macromol. Mater. Eng. **301**, 496–522 (2016)

37. G. Crini, L. Janus, M. Morcellet, G. Torri, N. Morin, Sorption properties toward substituted phenolic derivative in water using macroporous polyamines containing β-cyclodextrin. J. Appl. Polym. Sci. **73**, 2903–2910 (1999)

38. N. Sakairi, N. Nishi, S. Tokura (eds.), *Polysaccharide Applications* (1999)

39. P. Le-Thuaut, B. Martel, G. Crini, U. Maschke, X. Coqueret, M. Morcellet, Grafting of cyclodextrins onto polypropylene nonwoven fabrics for the manufacture of reactive filters. I. Synthesis parameters. J. Appl. Polym. Sci. **78**, 2118–2125 (2000)

40. H.L. Chen, L.G. Wu, J. Tan, C.L. Zhu, PVA membrane filled β-cyclodextrin for separation of isomeric xylenes by pervaporation. Chem. Eng. J. **78**, 159–164 (2000)

41. S.T. Lee, F.L. Mi, Y.J. Shen, S.S. Shyu, Equilibrium and kinetic studies of copper(II) ion uptake by chitosan–tripolyphosphate chelating resin. Polymer **42**, 1879–1892 (2001)

42. G.C. Steenkamp, K. Keizer, H.W.J.P. Neomagus, H.M. Krieg, Copper II removal from polluted water with alumina/chitosan composite membranes. J. Membr. Sci. **197**, 147–156 (2002)

43. X.D. Liu, S. Tokura, N. Nishi, N. Sakairi, A novel method for immobilization of chitosan onto non-porous glass beads through a 1,3-thiazolidine linker. Polymer **44**, 1021–1026 (2003)

44. X.D. Liu, S. Tokura, M. Haruki, N. Nishi, N. Sakairi, Surface modification of nonporous glass beads with chitosan and their adsorption property for transition metal ions. Carbohydr. Polym. **49**, 103–108 (2002)

45. Y. Fan, Y.Q. Feng, S.L. Da, On-line selective solid-phase extraction of 4-nitrophenol with β-cyclodextrin bonded silica. Anal. Chim. Acta **484**, 145–153 (2003)

46. N.A. Travlou, G.Z. Kyzas, N.K. Lazaridis, E.A. Deliyanni, Functionalization of graphite oxide with magnetic chitosan for the preparation of a nanocomposite dye. Langmuir **29**, 1657–1668 (2013)

47. L. Dehabadi, L.D. Wilson, Polysaccharide-based materials and their adsorption properties in aqueous solution. Carbohyd. Polym. **113**, 471–479 (2014)

48. F.J. Morales-Sanfrutos, M.A.A. Lopez-Jaramillo, F. Elremaily, F. Hernández-Mateo, F. Santoyo-Gonzalez, Divinyl sulfone cross-linked cyclodextrin-based polymeric materials: synthesis and applications as sorbents and encapsulating agents. Molecules **20**, 3565–3581 (2015)

49. L. Wang, A. Wang, Adsorption characteristics of Congo Red onto the chitosan/montmorillonite nanocomposite. J. Hazard. Mater. **147**, 979–985 (2007)

50. H.C. Lee, Y.G. Jeong, B.G. Min, W.S. Lyoo, S.C. Lee, Preparation and acid dye adsorption behavior of polyurethane/chitosan composite foams. Fibers Polym. **10**(5), 636–642 (2009)

51. M.Y. Chang, R.S. Juang, Adsorption of tannic acid, humic acid and dyes from water using the composite of chitosan and activated clay. J. Colloid Interface Sci. **278**, 18–25 (2004)

52. W.S.W. Ngah, L.C. Teong, M.A.K.M. Hanafiah, Adsorption of dyes and heavy metal ions by chitosan composites: a review. Carbohydr. Polym. **83**, 1446–1456 (2011)

53. L.S. Casey, L.D. Wilson, Investigation of chitosan-PVA composite films and their adsorption properties. J. Geosci. Environ. Prot. **3**, 78–84 (2015)
54. M.B. Veera, A. Krishnaiah, L.T. Jonathan, D.S. Edgar, H. Richard, Removal of arsenic (III) and arsenic (V) from aqueous medium using chitosan-coated biosorbent. Water Res. **42**, 633–642 (2008)
55. T.V. Budnyak, I.V. Pylypchuk, V.A. Tertykh, E.S. Yanovska, D. Kolodynska, Synthesis and adsorption properties of chitosan-silica nanocomposite prepared by sol-gel method. Nanoscale Res. Lett. **10**, 87 (2015)
56. X.X. Li, J. Li, X.J. Sun, L.Y. Cai, Y.C. Li, X. Tian, J.R. Li, Preparation and malachite green adsorption behavior of polyurethane/chitosan composite foam. J. Cell. Plast. **51**(4), 373–386 (2014)
57. K.A. Kristiansen, A. Potthast, B.E. Christensen, Periodate oxidation of polysaccharides for modification of chemical and physical properties. Carbohyd. Res. **345**(10), 1264–1271 (2010)
58. Y. Zhang, Z. Shen, C. Dai, X. Zhou, Removal of selected pharmaceuticals from aqueous solution using magnetic chitosan: sorption behavior and mechanism. Environ. Sci. Pollut. Res. **21**, 12780–12789 (2014)
59. T. Heinze, T. Liebert, Unconventional methods in cellulose functionalization. Prog. Polym. Sci. **26**, 1689–1762 (2001)
60. S.Y. Oh, D.I. Yoo, Y. Shin, H.C. Kin, H.Y. Kim, Y.S. Chuang, Crystalline structure analysis of cellulose treated with sodium hydroxide and carbon dioxide by means of X-ray diffraction and FTIR spectroscopy. Carbohyd. Res. **340**, 2376–2391 (2005)
61. K.S. Low, C.K. Lee, S.M. Mak, Sorption of copper and lead by citric acid modified wood. Wood Sci. Technol. **38**, 629–640 (2004)
62. M. Marchetti, A. Clement, B. Loubinoux, P. Gerardin, Decontamination of synthetic solutions containing heavy metals using chemically modified saw-dusts bearing polyacrylic acid chains. J. Wood Sci. **46**, 331–333 (2000)
63. N. Aoki, K. Fukushima, H. Kurukata, M. Sakamoto, K. Furuhata, 6-Deoxy6-mercaptocellulose and its S-substituted derivatives as sorbents for metal ions. React. Funct. Polym. **42**, 223–233 (1999)
64. S. Camy, S. Montanari, M. Vignon, J.S. Condores, Oxidation of cellulose on pressurized carbon dioxide. J. Supercrit. Fluids **51**, 188–196 (2009)
65. R.R. Navarro, K. Sumi, N. Fuji, M. Matsumura, Mercury removal from wastewater using porous cellulose carrier modified with polyethyleneimine. Water Res. **30**, 2488–2494 (1996)
66. X.Q. Sun, B. Peng, Y. Jing, J. Chen, D.Q. Li, Chitosan(chitin)/cellulose composite biosorbents prepared using ionic liquid for heavy metal ions adsorption. Separations **55**, 2062–2069 (2009)
67. N. Hameed, Q.P. Guo, Natural wool/cellulose acetate blends regenerated from the ionic liquid 1-butyl-3-methylimidazolium chloride. Carbohyd. Polym. **78**, 999–1004 (2009)
68. T. Ueki, M. Watanabe, Macromolecules in ionic liquids: progress, challenges and opportunities. Macromolecules **41**, 3739–3749 (2008)
69. Y. Jiang, W. Wang, X. Li, X. Wang, J. Zhou, X. Mu, Enzyme-mimetic catalyst-modified nanoporous SiO_2-cellulose hybrid composites with high specific surface area for rapid H_2O_2 detection. ACS Appl. Mater. Interfaces **5**, 1913–1916 (2013)
70. R. Soleyman, G.R. Bardajeeb, A. Pourjavadi, A. Varamesh, A.A. Davoodi, Hydrolyzed salep/gelatin-g-polyacrylamide as a novel micro/nano-porous superabsorbent hydrogel: synthesis, optimization and investigation on swelling behavior. Sci. Iranica **22**, 883–893 (2015)
71. K. Abe, S. Iwamoto, H. Yano, Obtaining cellulose nanofibres with a uniform width of 15 nm from wood. Biomacromolecules **8**, 3276–3278 (2007)
72. Y. Fan, T. Saito, A. Isogai, Chitin nanocrystals prepared by TEMPO-mediated oxidation of alpha-chitin. Biomacromolecules **9**, 192–198 (2008)

73. Q. Xu, W. Li, Z. Cheng, G. Yang, M. Qin, TEMPO/NaBr/NaClO-mediated surface oxidation of nanocrystalline cellulose and its microparticulate retention system with cationic polyacrylamide. BioResources **9**, 994–1006 (2014)

74. S. Maxwell, S. Yates, *The Future of Water* (American Water Works Association, 2011)

75. V.V. Goncharuk, *Prospects of Development of Fundamental and Applied investigations in the field of water physics, chemistry and biology* (Naukova dumka, Kyiv (Russian), 2011)

76. I. Simkovic, Review: what could be greener than composites made from polysaccharides? Carbohydr. Polym. **74**, 759–762 (2008)

77. A. Tiwari, *Polysaccharides: Development, Properties and Applications* (Nova Science Publisher Inc., New York, 2010)

78. M.A. Atieh, Removal of phenol from water different types of carbon—a comparative analysis. APCBEE Procedia **10**, 136–141 (2014)

79. R. Boopathy, M. Wilson, C.F. Kulpa, Anaerobic removal of 2,4,6-trinitrotoluene (TNT) under different electron accepting conditions: laboratory study. Water Environ. Res. **65**(3), 271–275 (1993)

80. E. Kachlishvili, M. Asatiani, A. Kobakhidze, V. Elisashvili, Trinitrotoluene and mandarin peels selectively affect lignin-modifying enzyme production in white-rot basidiomycetes. SpringerPlus **5**, 252–260 (2016)

81. C.I. Pearce, J.R. Lloyd, J.T. Guthrie, The removal of colour from textile wastewater using whole bacterial cells: a review. Dyes Pigm. **58**, 179–196 (2003)

82. G. Durai, M. Rajasimman, Biological treatment of tannery wastewater. J. Environ. Sci. Technol. **4** (2011)

83. J. Volker, S. Castronovo, A. Wick, T.A. Ternes, A. Joss, J. Oehlmann, M. Wagner, Advancing biological wastewater treatment: extended anaerobic conditions enhance the removal of endocrine and dioxin-like activities. Environ. Sci. Technol. **50**, 10606–10615 (2016). doi: 10.1021/acs.est.5b05732

84. B. Van der Bruggen, C. Vandecasteele, Removal of pollutants from surface water and groundwater by nanofiltration: overview of possible applications in the drinking water industry. Environ. Pollut. **122**, 435–445 (2003)

85. A. Lee, J.W. Elam, S.B. Darling, Membrane materials for water purification: design, development and application. Environ. Sci. Water Res. Technol. **2**, 17–42 (2016)

86. F. Al-Momani, E. Touraud, J.R. Degorce-Dumas, J. Roussy, O. Thomas, Biodegradability enhancement of textile dyes and textile wastewater by VUV photolysis. J. Photochem. Photobiol. A Chem. **153**, 191–197 (2002)

87. T. Kurbus, Y.M. Slokar, A. Majcen Le-Marechal, D.B. Voncina, The use of experimental design for the evaluation of the influence of variables on the H_2O_2/UV treatment of model textile waste water. Dyes Pigm. **58**, 171–178 (2003)

88. K.E. O'shea, D.D. Dionysiou, Advanced oxidation processes for water treatment. J. Phys. Chem. Lett. **3**, 2112–2113 (2012)

89. S. Giannakis, F.A.G. Vives, D. Grandjean, A. Magnet, L. De-Alencastro, C. Pulgarin, Effect of advanced oxidation processes on the micropollutants and the effluent organic matter contained in municipal wastewater previously treated by three different secondary methods. Water Res. **84**, 295 (2015)

90. B.P. Chaplin, Critical review of electrochemical advanced oxidation processes for water treatment applications. Environ. Sci. Process. Impacts **16**, 1182–1203 (2014)

91. M.X. Loukidou, K.A. Matis, A.I. Zouboulis, M. LiakopoulouKyriakidou, Removal of As(V) from wastewaters by chemically modified fungal biomass. Water Res. **37**, 4544–4552 (2003)

92. S. Netpradit, P. Thiravetyan, S. Towprayoon, Application of waste metal hydroxide sludge for adsorption of azo reactive dyes. Water Res. **38**, 71–78 (2004)

93. J. Akhtar, N. Aishah, S. Amin, K. Shahzad, A review on removal of pharmaceuticals from water by adsorption. Desalin. Water Treat. **57**, 12842–12860 (2015)

94. C. Sophia, E.C. Lima, N. Allaudeen, S. Rajan, Application of graphene based materials for adsorption of pharmaceutical traces from water and wastewater—a review. Desalin. Water Treat. (2016). doi:10.1080/19443994.2016.1172989

95. Y. Tsuchiya (ed.), Water Quality and Standards-Vol. II Inorganic Chemicals Including Radioactive Materials in Waterbodies, Encyclopedia of Life Support Systems (2000)
96. R.J. Qu, C.M. Sun, F. Ma, Y. Zhang, C. Ji, Q. Xu, C. Wang, H. Chen, Removal and recovery of Hg(II) from aqueous solution using chitosan-coated cotton fibers. J. Hazard. Mater. **167**, 717–727 (2009)
97. G.Y. Zhang, R.J. Qu, C.M. Sun, C.N. Ji, H. Chen, C.H. Wang, Y. Niu, Design and synthesis of multifunctional materials based on an ionic-liquid backbone. Angew. Chem. **118**, 5999–6002 (2008)
98. R.J. Qu, C.M. Sun, M.H. Wang, C.N. Ji, Q. Xu, Y. Zhang, Adsorption of Au(III) from aqueous solution using cotton fiber/chitosan composite adsorbents. Hydrometallurgy **100**, 65–71 (2009)
99. A. Anjum, C.K. Seth, M. Datta, Removal of As3+ using chitosan–montmorillonite composite: sorptive equilibrium and kinetics. Adsorpt. Sci. Technol. **323** (2013)
100. F.A. Pereira, K.S. Sousa, G.R. Cavalcanti, M.G. Fonseca, A.G. de Souza, A.P. Alves, Chitosan-montmorillonite biocomposite as an adsorbent for copper (II) cations from aqueous solutions. Int. J. Biol. Mol. **61**, 471–478 (2013)
101. S.I. Park, I.S. Kwak, S.W. Won, Y.S. Yun, Glutaraldehyde-crosslinked chitosan beads for sorptive separation of Au(III) and Pd(II): Opening a way to design reduction-coupled selectivity-tunable sorbents for separation of precious metals. J. Hazard. Mater. **248–249**, 211–218 (2013)
102. Y. Cheng, K. Xu, H. Li, Y. Li, B. Liang, Preparation of urea-imprinted cross-linked chitosan and its adsorption behavior. Anal. Lett. **47**, 1063–1078 (2014)
103. A. Pourjavadi, Z.M. Tehrani, H. Salimi, A. Banazadeh, N. Abedini, Hydrogel nanocomposite based on chitosan-g-acrylic acid and modified nanosilica with high adsorption capacity for heavy metal ion removal. Iran. Polym. J. **24**, 725–734 (2015)
104. J. Xie, C. Li, L. Chi, D. Wu, Chitosan modified zeolite as a versatile adsorbent for the removal of different pollutants from water. Fuel, **103**, 480–485 (2013)
105. B. Damià, *Emerging Organic Pollutants in Waste Waters and Sludge* (Springer, Berlin, 2005)
106. I.M. Ali, T.A. Khan, Low cost adsorbents for the removal of organic pollutants from Wastewater. J. Environ. Manage. **113**, 170–183 (2012)
107. S. Harrad, *Persistent Organic Pollutants Environmental Behaviour and Pathways for Human Exposure* (Kluwer Academic Publishers, Norwell, 2001)
108. L.P. Burkhard, M.T. Lukazewycz, Toxicity equivalency values for polychlorinated biphenyl mixtures. Environ. Toxicol. Chem. **27**, 529–534 (2008)
109. T. Clive, Persistent Organic Pollutants: are we close to a solution? Arctic Resources Committee, vol. 26, Number 1, Fall/Winter (2000)
110. H. Fiedler, *Persistent Organic Pollutants* (Springer, New York, 2003)
111. L. Ritter, K.R. Solomon, J. Forget, Persistent organic pollutants: an assessment report, in *Canadian Network of Toxicology Centres Guelphon* (2000)
112. L. Poon, L.D. Wilson, J.V. Headley, Chitosan-glutaraldehyde copolymers and their sorption properties. Carbohydr. Polym. **109**, 92–101 (2014)
113. L. Poon, S. Younus, L.D. Wilson, Adsorption study of an organo-arsenical with chitosan based sorbents. J. Colloid Interface Sci. **420**, 136–144 (2014)
114. D.Y. Pratt, L.D. Wilson, J.A. Kozinski, Preparation and sorption studies of glutaraldehyde cross-linked chitosan copolymers. J. Colloid Interface Sci. **395**, 205–211 (2013)
115. L.C. Zhou, X.G. Meng, J.W. Fu, Y.C. Yang, P. Yang, C. Mi, Highly efficient adsorption of chlorophenols onto chemically modified chitosan. Appl. Surf. Sci. **292**, 735–741 (2014)
116. G.Z. Kyzas, M. Kostoglou, N.K. Lazaridis, D.A. Lambropoulou, D.N. Bikiaris, Environmental friendly technology for the removal of pharmaceutical contaminants from wastewaters using modified chitosan adsorbents. Chem. Eng. J. **222**, 248–258 (2013)

117. G.Z. Kyzas, D.N. Bikiaris, M. Seredych, T.J. Bandosz, E.A. Deliyanni, Removal of dorzolamide from biomedical wastewaters with adsorption onto graphite oxide/poly(acrylic acid) grafted chitosan nanocomposite. Bioresour. Technol. **152**, 399–406 (2014)

118. H. Kono, Preparation and characterization of amphoteric cellulose hydrogels as adsorbents for the anionic dyes in aqueous solutions. Gels **1**, 94–116 (2015)

119. M.N.V.R. Kumar, A review of chitin and chitosan applications. React. Funct. Polym. **46**, 1–27 (2000)

120. A.J. Varma, S.V. Deshpande, J.F. Kennedy, Metal complexation by chitosan and its derivatives: a review. Carbohydr. Polym. **55**, 77–93 (2004)

121. F. Delval, G. Crini, N. Morin, J. Vebrel, S. Bertini, G. Torri, The sorption of several types of dye on crosslinked polysaccharides derivatives. Dyes Pigm. **53**, 79–92 (2002)

122. G. Crini, Studies on adsorption of dyes on beta-cyclodextrin polymer. Bioresour. Technol. **90**, 193–198 (2003)

123. T.N.T. Phan, M. Bacquet, M. Morcellet, Synthesis and characterization of silica gels functionalized with monochlorotriazinyl β-cyclodextrin and their sorption capacities towards organic compounds. J. Incl. Phenom. Macrocyl. Chem. **38**, 345–359 (2000)

124. T.N.T. Phan, M. Bacquet, M. Morcellet, The removal of organic pollutants from water using new silica-supported β-cyclodextrin derivatives. React. Funct. Polym. **52**, 117–125 (2002)

125. G. Crini, M. Bourdonneau, B. Martel, M. Piotto, M. Morcellet, T. Richert, Solid-state NMR characterization of cyclomaltoheptaose (β-cyclodextrin) polymers using high resolution magic angle spinning with gradients. J. Appl. Polym. Sci. **75**, 1288–1295 (2000)

126. S. Lata, P.K. Singh, Regeneration of adsorbents and recovery of heavy metals: a review. Int. J. Environ. Sci. Technol. **12**, 1461–1478 (2015)

127. M.O. Omorogie, J.O. Babalola, E.I. Unuabonah, Regeneration strategies for spent solid matrices used in adsorption of organic pollutants from surface water: a critical review. Desalination and water treatment. **57**(2), 518–544 (2016)

128. X. Luo, X. Lei, N. Cai, X. Xie, Y. Xue, F. Yu, Removal of heavy metal ions from water by magnetic cellulose-based beads with embedded chemically modified agnetite nanoparticles and activated carbon. ACS Sustain. Chem. Eng. **4**, 3960–3969 (2016)

129. T.V.J. Charpentier, A. Neville, J.L. Lanigan, R. Barker, M.J. Smith, T. Richardson, Preparation of magnetic carboxymethylchitosan nanoparticles for adsorption of heavy metal ions. ACS Omega **1**, 77–83 (2016)

130. M. Onditi, A.A. Adelodun, E.O. Changamu, J.C. Ngila, Removal of Pb2+ and Cd2+ from drinking water using polysaccharide extract isolated from cactus pads (Opuntia ficus indica). J. Appl. Polym. Sci. (2016). doi:10.1002/APP.43913

131. I. Mallard, L. W. Staede, S. Ruellan, P.A.L. Jacobsen, K.L. Larsen, S. Fourmentin, Synthesis, characterization and sorption capacities toward organic pollutants of new β-cyclodextrin modified zeolite derivatives. Colloids and Surfaces A: Physicochem. Eng. Aspects **482**, 50–57 (2015)

132. T. Gotoh, K. Matsushima, K.I. Kikuchi, Preparation of alginate–chitosan hybrid gel beads and adsorption of divalent metal ions. Chemosphere **55**, 135–140 (2004)

133. R.W. Coughlin, M.R. Deshaies, E.M. Davis, Chitosan in crab shell wastes purifies electroplating wastewater. Environ. Prog. **9**, 35–39 (1990)

134. W.S. Ngah, I.M. Isa, Comparison study of copper ion adsorption on chitosan, Dowex A-1, and Zerolit 225. J. Appl. Polym. Sci. **67**, 1067–1070 (1998)

135. B. Martel, M. Devassine, G. Crini, M. Weltrowski, M. Bourdonneau, M. Morcellet, Preparation and sorption properties of a β-cyclodextrin-linked chitosan derivative. J. Appl. Polym. Sci. **39**, 169–176 (2001)

136. C.P. Okoli, G.O. Adewuyi, Q. Zhang, Q. Guo, QSAR aided design and development of biopolymer-based SPE phase for liquid chromatographic analysis of polycyclic aromatic hydrocarbons in environmental water samples. RSC Adv. (2016) doi:10.1039/C6RA10932B

137. E.G. Furuya, H.T. Chang, Y. Miura, K.E. Noll, A fundamental analysis of the isotherm for the adsorption of phenolic compounds on activated carbon. Sep. Purif. Technol. **11**, 69–78 (1997)

138. G. Crini, S. Bertini, G. Torri, A.M. Naggi, D. Sforzini, C. Vecchi, Sorption of aromatic compounds in water using insoluble cyclodextrin polymers. J. Appl. Polym. Sci. **68**, 1973–1978 (1998)
139. S. Babel, T.A. Kurniawan, Low-cost adsorbents for heavy metals uptake from contaminated water: a review. J Hazard. Mater. **97**, 219–243 (2003)
140. K. Kurita, Controlled functionalization of the polysaccharide chitin. Prog. Polym. Sci. **26**, 1921–1971 (2001)

Chapter 5
Tapping into Microbial Polysaccharides for Water and Wastewater Purifications

5.1 Introduction

Wastewater treatment systems have contributed largely to the availability of water as water reuse is greatly advocated globally. Wastewater treatment can be grouped into three categories i.e. mechanical, aquatic, and terrestrial, and into three phases i.e. primary, secondary and tertiary treatments. With these treatment systems, a common problem of microbial clogging, slime bulking and biofouling characterized by the prevalence of biofilms formed from exopolysaccharides (EPS) and other polymeric substances have been identified. Polysaccharides can thus be utilized in the efficient treatment of wastewater, by enhancing the water treatment processes (Fig. 5.1).

Microbial polysaccharides are one of the vast ranges of extracellular biopolymers produced by microorganisms through the utilization of simple to complex substrates [1]. They show considerable diversity in their composition and structure. In nature, they play crucial roles in maintaining the cell viability by conserving genetic information, producing energy or reducing power, protecting microbes from invasion and storing of carbon-based macromolecules [2]. The recent interest in exploiting renewable sources as alternatives for synthetic chemicals has brought the exploration of microbial polysaccharides and their full recommendation and usage [3]. The reasons for this interest in microbial polysaccharides are that they are non-toxic and biodegradable polymers found on the surfaces of microbial cells, playing varying roles in the biological activities of the cell [4].

They are large molecular weight carbohydrates polymers with diverse biological functions, which could exist as an attachment on the outer cell membrane as lipopolysaccharides (LPS), a secretion with discrete surface layers (Capsular polysaccharides) or as excretions which are loosely connected to the cell wall (exopolysaccharides) [4, 5]. Of these three unique forms, the exopolysaccharides (EPS) represent a multifunctional class, i.e. they are of higher biological functions. Although with several functional roles, microbial polysaccharides still require

© The Author(s) 2017
N.A. Oladoja et al., *Polysaccharides as a Green and Sustainable Resource for Water and Wastewater Treatment*, Biobased Polymers,
DOI 10.1007/978-3-319-56599-6_5

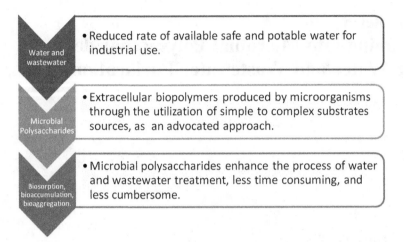

Fig. 5.1 Microbial polysaccharides and wastewater treatment

chemical modification to be acceptable for full utilization. Their applications are quite diverse ranging from agriculture, nanotechnology, biotechnology, health to cosmetic industries, etc. [4]. Presently, their usefulness as biofilms in wastewater treatments has been quite revealing with numerous studies exploring their uses in this regard.

Microbial polysaccharides secreted into the environment are referred to as exopolysaccharides (EPS). These EPS could be classified based on the composition of the monosaccharide present in them; while homo-polysaccharides have only one type of monosaccharide present, hetero-polysaccharides have more than one type of monosaccharide present [6].

Microbial polysaccharides have found direct usefulness in wastewater treatments. Their usage as replacements for chemical coagulants in the coagulation-flocculation procedure in water treatments have even become more relevant in recent years due to the health and environmental challenges posed by chemical coagulants [7]. Microbial polysaccharide usefulness has also been demonstrated in biological filtration systems. Biofilms formed as a result of these polysaccharides acts as bio-filters which are most suitable than other filtration systems in removing organic contaminants. They are environmentally safe, offering little or no negative health implications.

5.2   General Applications of Microbial Polysaccharides

Microbial polysaccharides have found extensive usage in different spheres. They have been employed in food industries with different functions such as utilizing them as viscosifying agents, stabilizers, emulsifiers, gelling agents, or water-binding agents in food [8]. Although lots of the EPS used in the food industries are from the plant [8], microbial polysaccharides have been found to be very useful in this particular industry, due to their viscous nature even at low concentrations and their pseudoplastic nature [9]. For instance, xantham which is made from *Xanthomonas campestris* is useful as a food packaging material, as well as a viscosifier, stabilizer, emulsifier and suspending agent in food industries. Also, xylinan made from *Acetobacter xylinum* is used in the food industry as a viscosifier and gelling agent while *levan* is used in the production of sweet confectionary and Gellan as a stabilizer and suspending agents for foods [10, 11].

Bacteria hyaluronic acid (HA) has continually shown great usage in biomedical applications. They are used in viscosurgery, serving as a replacer of eye fluid in ophthalmic surgery [10, 11]. They are also used in cosmetic surgery (viscoaugmentation), synovial fluid replication (viscosupplementation), in wound healing (viscoseparation), etc. Dextran, have been used to prepare plasma substitutes used during blood loss [1]. Microbial polysaccharides have also been shown to possess cholesterol-lowering abilities, hence their use as probiotics. Capsular polysaccharide produced by gut bacteria promotes the ability of micriobial polysaccharides to adhere to surfaces [12]. In a study by Tok and Aslim [13], it was shown that strains of *Lactobacillus delbrueckii* producing high amount of EPS removed more cholesterol from the experimental media, compared to those producing a lower amount of EPS (Fig. 5.2).

Microbial polysaccharides have also been used as emulsifiers and biosurfactant (which are heteropolysaccharides) with huge applications in bioremediation. Their usage as these products is due to their low toxicity, biodegradable ability, production from renewable sources and stability under extreme conditions [8, 14]. A study by Jain et al. [14] showed that biosurfactant produced by *Cronobacter sakazaki* exhibited significant activity with oil (by enhancing their water solubility

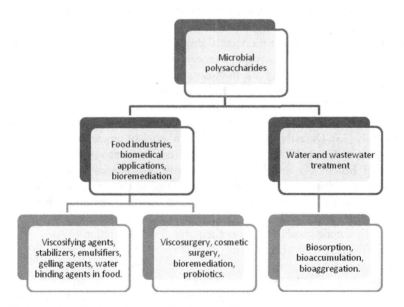

Fig. 5.2 General applications of microbial polysaccharides

and displacement from soil) [15] and hydrocarbon. *Pseudoalteromonas sp.* has been reported to exhibit high emulsifying activity against different ranges of oil products. Other bacteria known to produce biosurfactant include *Pseudomonas aeruginosa, Torulopsis bombicola, Bacillus subtilis* and *Candida* [16, 17].

5.3   Microbial Polysaccharide in Water and Wastewater Purification

Microbial polysaccharides biomasses are major resources in the purification of wastewaters of different origins. Wastewater treatment is usually a very expensive and time-consuming process. Microbial polysaccharides have shown that they possess great ability in making this cumbersome process less time-consuming and more efficient.

Biosorption is one of the major aspects microbial polysaccharides could be applied in water and wastewater treatment. Biosorption involves the removal of substances from solutions using microbial mass (chitin and polysaccharides being their principal components). Metals such as chromium, arsenic, mercury, aluminium, lead, cadmium, are common water pollutants that pose serious health risk to both humans and aquatic biota [18]. Hence, their removal from water and wastewater is of immense importance. Although various methods have been employed in removing heavy metals from water including chemical precipitation, ion exchange, coagulation-flocculation, electro-wining, and electro-coagulation [19, 20], biosorption has been used extensively for the removal of metals (including heavy metals) during the treatment of water and wastewater.

In biosorption, microbial biomass use their chemical functionalities to concentrate these metals, making their removal easier. The biosorption process involves sorption processes in which the metals are concentrated and accumulated in the microbial biomass [21] and desorption processes which cater for the recovery of the microbial biomass for reuse [22]. In biosorption, metal ions (which are positively charged) could be absorbed by binding to microbial cell surfaces which are negatively charged. This is because the prevalence of microbial polysaccharides in microbial cell walls in the form of EPS, makes available abundant binding sites for these metals [23]. Once this binding occurs in a series of processes which is unique for different metals and microorganisms involved, the removal of the metals from wastewater becomes a straight forward process. Advantages of using microbial polysaccharides (biomass) in biosorption compared to other methods of removing metals from water include; being growth independent, as non-living biomass are not affected by toxicity [23], low cost, regeneration of biosorbent and high efficiency [24], aseptic conditions are not required for its operation [20].

Textile industry wastewater remain one of the most difficult wastewater to treat not just because of the prevalence of heavy metals and aromatic hydrocarbons, but also due to the presence of large organic molecules called dyes. Adsorption is a crucial water and wastewater treatment process. It has been shown to be an effective procedure in treating wastewaters of industrial origin. Adsorption has been referred to as "one of the most promising decolourization techniques in dyeing wastewater treatment" [25]. Microbial polysaccharides (in the form of biomass) have been used as good biosorbents in adsorbing dyes during wastewater treatment. Polysaccharide-based materials are shown to possess numerous advantages in the removal of pollutants from wastewaters [26]. In adsorption, biosorbents produced from microbial polysaccharides acts as surfaces for the attachment of dyes and ions from metals, after accumulation before eventual removal. Adsorption works more like adhesion with attractions usually in the form of van der Waals forces physisorption and/or chemisorption. Dye removal from aqueous solution involves two processes of adsorption and ion exchange, with its application employed for either dye removal alone, or a combination of dye removal and water and wastewater treatment [25].

Another importance of microbial polysaccharides in wastewater treatment is in bioaggregation. Bioaggregation is a major required characteristic of microbial polysaccharides when utilized for wastewater treatment. This particular feature is what influences the flocculability, settleability, and dewaterability for flocs and sludge retention and shear resistance for biofilms [27].

Bioaccumulation is another process microbes are used in wastewater treatment. In Bioaccumulation, microbes accumulate metals in their biomass effectively, thus when this biomass are removed from the water a large proportion of the metal which they accumulated in their biomass would have been removed along with it. This is quite common with phosphate removing bacteria, generally referred to as polyphosphate accumulating organisms, achieving phosphorus removal through intracellular accumulation of their biomass [28].

5.4 Characterization of Microbial Polysaccharides for Water and Wastewater Purification

Various microorganisms have shown their high efficiency in wastewater treatments. Microbial polysaccharides are part of the complex EPS present in biofilms used in wastewater treatment. Thus characterization of microbial polysaccharides and biofilms is done to understand the structural diversity of these polysaccharides and their applicability in different industries. It also improves our understanding of how they bind to metals, act as adsorbents and possess other characteristics which they exhibit in great proportion.

Microbial polysaccharides are a part of extracellular polymeric substances produced by microbes. These substances apart from polysaccharides contain proteins, lipids, nucleic acids [29]. Thus characterization of microbial polysaccharides is done after their production, extraction and isolation. Various methods have been employed in the characterization of microbial polysaccharides. The standard but effective methods include Thin Layer Chromatography (TLC), Gas Chromatography (GC), Gas Chromatography Mass Spectrometry (GC-MS), High Performance Liquid

Chromatography (HPLC). Kassim [30], was able to show that the constituents of xanthan gum produced from *Xanthomonas campestris* consisted of glucose, glucouronic acid, mannose, acetic and pyruvic acids using thin layer chromatography.

Irrespective of the methods used in the characterization of microbial polysaccharides, its efficiency depends on the amount of the polysaccharide available which in turn is dependent on medium, culture production process and type of EPS. Carbohydrate components of a culture media have been known to affect the production of EPSs with necessarily influencing their chemical structures [31].

In the characterization of microbial polysaccharides, a necessary step is to first determine the monosaccharide components of these polysaccharides [31], as the overall characteristics and functional properties exhibited by these microbial polysaccharides are mainly dependent on these individual components. These processes are usually referred to as depolymerisation, and can be achieved using acid hydrolysis as described by Madhuri and Prabhakar [6], were trifluoroacetic acid (TFA), formic acid, sulphuric acid, or hydrochloric acid are used at 100 °C to hydrolyze the glycosidic bonds between the monomer molecules

High Performance Liquid Chromatography (HPLC) with UV detector is one of the most common analytical methods used for the characterization of microbial polysaccharides; this is due to its ability to carry out both qualitative and quantitative analysis of the polysaccharides. Several analytical methodologies have been successfully established and used for microbial characterization [32–34]. Gas Chromatography is another common tool used in the characterization of microbial polysaccharides, due to its high accuracy.

More recently, specialized probe-based scanning technologies, have been developed and used in the characterization of microbial polysaccharides. These systems are usually faster, and more efficient. One of such system developed is the Atomic Force Microscopy (AFM) which functions by measuring the force between a probe and a sample, producing high-resolution images of the structure under investigation [35]. Su et al. [36] used AFM to characterize polysaccharide capsules from *Zunongwangia profunda*. Their study showed that AFM polysaccharide capsules could be detected even in the presence of water covering capsule. Nuclear magnetic resonance and nonmetric multidimensional scaling are other approaches which have been used in the characterization of microbial polysaccharides [37].

5.5   Morphologic and Functional Properties of Microbial Polysaccharides

Microbial polysaccharides are categorized morphologically into the three unique forms aforementioned; Lipopolysaccharides (LPS), Capsular polysaccharides (CPS) and Exopolysaccharides (EPS).

5.5.1 Lipopolysaccharides (LPS)

Lipopolysaccharides are composed of moieties of macromolecules held together by hydrophobic forces which are generated from non-polar lipids found in the inner portion of the bilayers. The phospholipid bilayers also possess ionic interactions, the divalent cations Ca^{2+} and Mg^{2+} with which LPS binds to the cell wall structure. The chelating ionic interactions confer lateral stability on the outer membrane of the cell. Ligands for bacterial attachment and resilience to phagocytosis are provided by the O-specific polysaccharide region of the LPS [1]. The LPS functions as a permeability barrier, adhesins (adhesion receptors) for colonization, antigenic determinants of a strain and they mediate the release of cellular components resulting from metabolic components in biofilm formation. The structure of the LPS can significantly influence water-binding capacity at the cell surface. LPS are common features of Gram-negative bacteria.

5.5.2 Capsular Polysaccharides (CPS)

Capcular polysaccharides are an assembly of cohesive layers of polysaccharides bound together by covalent bonds, found on the outermost covering of a microbe [4]. They may often be highly immunogenic. They are also a common feature of bacteria and can be likened to the envelope of fungi. CPS functions in the following areas: aids the adherence of micriobial polysaccharides to surfaces, provides resistance against invasion to the cell, serves as a deterrent to desiccation due to their high affinity for water.

5.5.3 Exopolysaccharides (EPS)

Exopolysaccharides are the most diverse polysaccharides. They are synthesized and secreted into the environment by the cell wall-anchored enzymes. EPS are high molecular weight water-soluble polymers composed of sugar residues which may be neutral (non-ionic) or acidic (possess ionized groups such as the carboxyl) in nature [38]. EPS are composed of both hydrophobic and hydrophilic regions held together by ionic forces in a matrix. Microbial exopolysaccharides can be divided into two groups based on their monomeric composition: homo-polysaccharides and hetero-polysaccharides [4, 39]. Homo-polysaccharides are composed of a single monosaccharide unit while hetero-polysaccharides are composed of regularly repeating units of two to eight monosaccharide units. Categorization of EPS is very complex; hence further characterization factors such as the linkage bonds, nature of the monomeric units and types of monomeric units, are applied in creating a distinction between groups of EPS. The linkage bonds between the monomeric units: 1,4-β- or 1,3-β-linkages and 1,2-α- or 1,6-α-linkages, confer strong rigidity or flexibility, respectively on EPS [1]. The major attributes that make EPS functionally vast are biodegradability, biocompatibility, non-toxicity to humans and environment, edibility and rheological properties present [38]. In water, EPS due to their unique solubility and rheological properties are capable of dispersing, thickening or manipulating the viscosity effects, establishing its use in water and oil industries. Other physiological properties exhibited by EPS constituting its use in wastewater treatment include stabilization, synaeresis inhibition and suspension of particulates, crystallization control, encapsulation and formation of biofilm [5]. Also of extreme interest is the use of EPS as bioflocculants in water and wastewater treatment. Due to the high level of detrimental effects (carcinogenicity, Alzheimer's disease, neurotoxicity) caused by the use of synthetic polymers such as aluminium salts of the polyacrylamide derivatives as flocculants, there is need for an ecologically pure and safe alternative which can be provided by microbial EPS [40].

Exopolysaccharides play a crucial role in the world of biofilm. They constitute 50–90% of biofilm aggregates, but they vary in quantity depending on chemical and physical properties. They are involved in initiation of the adhesion process in colonization of surfaces, aggregation and immobilization of cells which increases the cell density of biofilm, water retention by the hydrophilic region, supplying nutrients, protection against external disruptions, adsorption and absorption of organic compounds and inorganic ions and storage of excess carbon used as energy in biofilm formation [4, 41]. EPS are produced by bacteria and fungi, and they

Fig. 5.3 Examples of commonly used polysaccharides

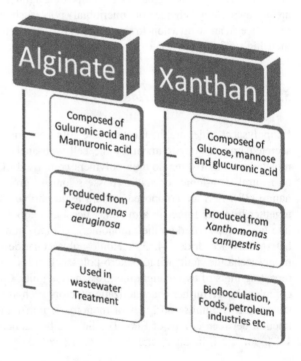

possess more stable and unique properties than Polysaccharides of algae and plants which are also used commercially [38].

Examples of commercially explored microbial EPS are Xanthan, Curdlan, Alginate, Cellulose, Dextran, Gellan, Hyaluronan, Levan, Pullulan, Scleroglucan, Succinoglycan and Glucoronan. Although with interesting functionalities, the high cost of production of microbial exopolysaccharides commercially is a constraint to its mass usage. Their production is also subject to physical variables such as pH, temperature and carbon source concentration [42] (Fig. 5.3).

5.6 Roles of Exopolysaccharides in Biofilm Formation

Microbial polysaccharides also known as biopolymers are self-generated extracellular polymeric substances with diverse functional properties produced by microbes. The most complex and versatile microbial polysaccharide is the Exopolysaccharides (EPS), with vast functional properties and impact in fields such as medical, pharmaceutical, bioremediation, water treatment and surgical applications. Examples include Xanthan, Glucan, Alginate, Curdlan and Gellan.

Biofilm refers to the communities of microorganisms embedded in a consortium of extracellular polymeric substances which develop on or attach to surfaces of animate or inanimate things. Biofilm formation serves as a prerequisite for the prolonged survival of almost 90% of microorganisms [43].

Exopolysaccharides (EPS) are necessary for the attachment of bacteria to surfaces during the development of biofilms [44] and they are a necessary factor in the colonization of bacteria on any surface. The biofilm matrix is composed of water, up to 97% [45], which constitutes the liquid and viscous phase and also the exopolysaccharides (50–60%) which confer mechanical stability on the viscoelastic properties observed in biofilms [46].

Aside exopolysaccharides, biofilm also contains nucleic acids, lipids, proteins and organic substances. Biofilms are said to be polymer gels with rheological properties and high porosity [46], which develop on surfaces where the attachment is possible, and there is an adequate flow of nutrients.

They can be found on medical devices, body tissues, aquatic and terrestrial environments, potable and industrial water pipes [47]. The major role of biofilm to microbial communities embedded within it is to protect them from environmental stress and external interference such as antibiotics, chemicals, bacteriophage and host immune responses. There is a difference in the behavior of microorganisms in biofilm and their planktonic nature, and the transition from planktonic growth to biofilm involves a multifaceted regulatory channel which sends signal for response to environmental changes and also acquisition and expression of genes required to mediate the reorganization of the microbial cells [48].

The roles of exopolysaccharides in biofilms involve; filling intracellular spaces present between bacteria, allowing the formation of biofilm matrix, protecting the bacteria cell walls from attack, desiccation, surviving under extreme conditions, adherence to surfaces, and increase in pathogenicity [49].

The formation of biofilm involves five phases:

In the first phase, a free-living microorganism attaches to a surface by van der Waal's forces. This phase is, however, reversible due to the weak attributes of van der Waal's forces. This progresses to the second phase in which the microbe adheres more strongly to the surface through the use of appendages (such as pilus) extending from the cell surface. The third phase exhibits the formation of the biofilm through cell division of the attached cells and also the attraction of other free-living microorganisms in the surrounding environments to the surface. The fourth phase is characterized by the secretion of exopolysaccharides by the microbial cells which forms a three-dimensional matrix. Lastly, the fifth phase involves the dispersal of the biofilm after climax has been reached and the biofilm can no longer hold together as a colony [50]. The most representative microbes in biofilm formation are bacteria. Biofilms are very resilient in nature; they adjust to fit into changes in environmental factors. Despite its negative significance in creating medical problems of outright antibiotic resistance, its advantage in water and wastewater treatment is undeniable due to its self-immobilizing, self-regenerating and biocatalytic properties which aid in water decontamination [51].

Antibiofilms are substances with the capability to disperse or inhibit microbial biofilms [52]. These substances can disrupt the activities of the bacterial cells and their attachments [53]. A couple of antibiofilm strategies have been tested, established and are in use. However, there is still continuous search for more ecologically pure and effective antibiofilm agents.

5.7 Microbial Biofilms in Water and Wastewater Treatment: The Dual Action

The uniqueness of microbial biofilms comes into play in water and wastewater treatment. They serve dual roles; negative and positive. Biofilms are present in water piping systems as recalcitrant contaminants because the exopolysaccharides present in the biofilm matrix constitute a permeability barrier against any toxic compounds or solvents that could harm the cells within. They cause corrosion in piping systems, clog pipes and are quite difficult to remove due to their high adhesive properties [51]. Enteric pathogenic microorganisms such as *Helicobacter pylori* may also be harboured by biofilms present in potable water distribution systems [54].

In water and wastewater treatment, the viscoelastic, as well as the resistant properties of biofilms, have been manipulated to be of advantageous use in oxidation of organic compounds and nitrification of ammonium [55]. A study carried out to test the effect of heavy metals on the planktonic and biofilm forms of *Pseudomonas aeruginosa* using a rotator disk biofilm reactor showed that the latter is 600 times more resistant to heavy metal stress [56]. Hence biofilms are not prone to environmental stress. Multi-species biofilms (biofilms containing different single species) are industrially and environmentally important, and they are majorly used in water and wastewater treatment. Studies have shown their effectiveness in removing organic compounds and heavy metals (especially Ammonia) in wastewater, hence preventing eutrophication which may arise from the deposition of these heavy metals into the environment [55].

Wastewater treatment systems such as trickling filter, modified lagoon used in the secondary treatment of waste employ biofilms in nutrient or waste removal. The biofilms are formed through the utilization of the dissolved nutrients present in the wastewater which consequently leads to the extrication of the nutrients [54]. The functional roles of biofilms in water and wastewater treatment include the following:

- Biofilms are capable of oxidizing organic constituents in wastewater, hence reducing the total organic compound present.
- Certain nitrifying microorganisms are embedded in biofilms. These microorganisms can oxidize ammonium to nitrite which is further oxidized to nitrates, therefore reducing ammonium concentration in wastewater.
- Biofilms effectively reduce the Chemical Oxygen Demand (COD) and the Biological Oxygen Demand (BOD). The reduction in the oxidizable matter in water results in a decreased need for oxygen and a subsequent reduction in COD.
- Biofilms also increase the level of nitrate in water through the oxidation of ammonium. The nitrate generated is also adsorbed by the biofilms as nutrients for active functioning.

Although the treatment of water and wastewater by biofilms could be less speedy, it is very effective at oxidizing organic compounds and ammonium. It is also relatively ecologically friendly. The advantages in the use of biofilms in wastewater treatment include flexibility in operation, space minimization, reduced hydraulic retention time, resilience to environmental stress, enhanced ability to degrade recalcitrant compounds and lower sludge production due to a slower microbial growth rate.

5.7.1   Antibiofilms in Water and Wastewater Treatment

A range of chemicals is used in removing biofilms from water and wastewater. However, the effectiveness of these chemicals depends on the concentration, temperature and pH of the water, the microbial load and the exposure time [57, 58]. Antibiofilms act in either of the following ways: removal of the biofilm matrix, destroying the microorganisms or preventing the biofilm formation. The chemical approach includes the use of chlorine, acids, sodium hypochlorite, phenolic, iodine

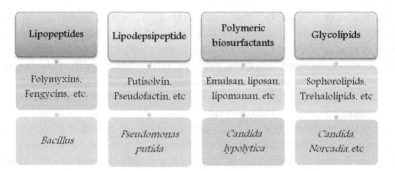

Fig. 5.4 Some classes of biosurfactants

complex and hydrogen peroxide [58]. The accumulation of the residues of these toxic chemicals in water could cause health issues and allergies; hence an ecologically safe approach is urgently required.

In recent years, the use of biosurfactants has been encouraged in water treatment as an eco-friendly approach and an alternative to complex conventional methods. Biosurfactants are heterogenous amphiphilic compounds of microbial origin with hydrophobic and hydrophilic regions secreted by microorganisms as a result of metabolism [17]. The unique properties of biosurfactants include low toxicity, biodegradability, surface and interfacial tension reduction, temperature and pH tolerance, emulsifying and de-emulsifying properties, anti-adhesive properties as well as antimicrobial properties. Biosurfactants are capable of inhibiting the formation of biofilms, disrupt the attachment and quorum sensing and detach the biofilms from surfaces [49, 57]. Biosurfactants are categorized by their microbial origin and chemical composition [57].

The constraints still experienced in the use of biosurfactants are a lack of vast information on the level of toxicity and due to the high cost of production. The current biological treatment methods of water and wastewater have proven to be ecologically friendly, safe and flexible with high prospects of efficacy if properly utilized. Figure 5.4 show the classes of biosurfactants and examples of each which can be utilized in the treatment of water and wastewater.

5.8 Analytical Methods for Assessment of Microbial Polysaccharides in Water and Wastewater

The analytical methods for assessment of microbial polysaccharides in water and wastewater during purification are usually the basic treatment procedures for industrial and potable water supplies. The objectives of these methods are to remove settleable organic and inorganic matter, residual organic and suspended solids, removal of heavy metals and other solids present in water and wastewater [44]. These methods are used to either double check the efficiency of a water treatment procedure on the removal of microbial polysaccharides (biofilm) or to measure the efficacy of the use of the microbial polysaccharides in wastewater treatment [59].

The parameters measured in the use of analytical methods for assessment of microbial polysaccharides include the following as shown in Fig. 5.5:

- Biological Oxygen Demand (BOD)
- Chemical Oxygen Demand (COD)
- Total suspended solids
- Nitrification/Denitrification.

5.8.1 Biological Oxygen Demand (BOD)

Biological Oxygen Demand (BOD) refers to the amount of oxygen required to degrade organic matter present in wastewater [60]. This depends on the source of the wastewater which could contain diverse organic matter such as food particles, detergents, faecal matter, oils and grease which are easily degraded by the microbial populace of the wastewater in the presence of oxygen. The microbial communities which make up the biofilm use the organic matter as a source of food, therefore reducing the amount of organic matter in the water [47]. The measurement of BOD is very crucial before releasing effluents from wastewater treatment into the environment. This is because a high amount of BOD in water released into the environment can cause eutrophication and changes in the ecosystem which may favour the excessive growth of microbes in the water. The active functioning of nitrifying and denitrifying bacteria is also influenced by the presence and level of BOD. The standard limit for BOD in potable water has been set at 30 mg BOD/L safe level [47]. Thus, it is theoretically accepted that the presence of a high concentration of BOD in wastewater indicates the presence of high organic matter to be degraded [47].

The conventional method of measuring BOD in water is the five-day BOD or BOD5 test, which involves the measurement of the amount of oxygen consumed by microorganisms (dissolved oxygen) in a five-day period. This shows the strength or the concentration of biodegradable organic matter present in the wastewater. The standard oxidation test time for BOD_5 has been set as 5 days at 20 °C temperature. Researchers have argued that the stipulated time for the BOD_5 test is too low to effectively quantify the amount of BOD; however, it has constantly been used to evaluate water quality for any immediate concern. The correct analysis of BOD can be interfered with by the presence of some constituents in the water sample to be analyzed. These constituents include the presence of toxic metals such as lead, presence of residual chlorine and caustic acidity or alkalinity, which can inhibit biochemical oxidation [60].

5.8.2 Chemical Oxygen Demand (COD)

Chemical Oxygen Demand (COD) is referred to as the relative amount of chemicals which consume dissolved oxygen in water. In water, both the organic and inorganic constituents are subject to oxidation. However, the oxidation of the organic constituents predominates. The measurement of COD in water is highly important because it represents the amount of oxygen consumed for aerobic biological oxidation or conversion of the organic matter in the water sample. A common procedure for determining the COD is the redox titration which involves heating the water sample with potassium dichromate; the organic matter in the water becomes oxidized to carbon dioxide and water while the dichromate is reduced to Chromic ion (Cr^{3+}). The amount of oxygen expended in converting organic matter to carbon dioxide and water is the COD. Therefore, COD value is expected to be low to ascertain the efficacy of treatment using microbial polysaccharides.

COD analysis and values can be interfered with by the digestion time, the concentration of the reagent, the presence of inorganic reducing compounds (nitrites, sulphites, iron (II) salts) that react quickly with dichromate ion thereby reducing its concentration, presence of oxidizing agents (such as chlorine, ammonia and amines) which may cause erroneously high COD values and the temperature of the water or wastewater.

5.8.3 Total Suspended Solids (TSS)

Total suspended solids (TSS) are all particles which are mostly inorganic, suspended in water. It is a measurement of solid material per volume of water. These particles are usually larger than 2 microns and therefore will not pass through the filter. Microbes such as bacteria and algae may also be constituents of the total solids because they form a mass (biofilm) in water, but may also function in the removal of the suspended solids. Other constituents of wastewater include sand, sediment and silt from erosion or other natural and anthropogenic activities. Sediments and silt usually do not occur in water in Free State; they have microorganisms which could be pathogenic adhering to them. TSS in water causes turbidity and makes the water unsafe for consumption by humans and aquatic life. TSS is expected to decrease through treatment of wastewater with microbial polysaccharides due to their adhesive properties.

TSS has usually measured accurately by filtering a water sample and weighing the residue retained on the filter. The Water Quality Standards posits that potable or safe water must contain less than 20 mg/L amount of total suspended solids.

Fig. 5.5 Indicators for monitoring microbial polysaccharides in water and wastewater treatment

5.8.4   Nitrification and Denitrification (Ammonium Test)

Wastewater contains nitrogen in many forms which include organic nitrogen produced by human materials (such as dead cells, urea) and inorganic nitrogen from
industrial wastes. Nitrogen exist in water in the form of Nitrates and Nitrites and are
only permissible in drinking water at a level not more than 10 mg/L for Nitrate and
1 mg/L for Nitrite [61]. The nitrogen present in wastewater can be degraded
completely to ammonia by the microbes in the water. Ammonia is further oxidized
to nitrite by the nitrifying bacteria, *Nitrosomonas* and *Nitrobacter,* in a process
known as Nitrification [62]. The nitrite is further oxidized to nitrate by the nitrifying
bacteria. However, the presence of excess nitrate in water is detrimental to human
and aquatic life as well as the environment as it may lead to eutrophication [61].
The second phase involves the conversion of nitrate to nitrogen gas in a process
called *Denitrification*. It is a theoretical expectation that wastewater being treated by
microbial polysaccharides should contain low ammonium concentration [62].

The method used in quantifying ammonium concentration in wastewater is
known as Nesslerization, which involves observing the colour change of the
mixture of the water sample and Nessler reagent [54].

Other heavy metals such as Lead, Phosphorus present in wastewater are also
absorbed by the microbial communities present. The absorption of these metals
results in the eventual degradation into soluble and non-toxic constituents. For
instance, phosphorus in wastewater is degraded to the non-toxic phosphates. The
removal of these metals is important because they are highly toxic to human health,
aquatic life and the environment at large.

The efficacy of microbial polysaccharides in water and wastewater treatment can be measured through the aforementioned analytical methods.

5.9 Trends and Prospects of Microbial Polysaccharides in Water and Wastewater Purification

Water and wastewater treatment systems need to be continuously modified and improved to meet the water requirements of most countries especially developing countries where water shortages is an issue of great concern. However, advancements in microbial polysaccharide based water treatment systems have bridged some of the gaps in solving problems associated with wastewater treatment.

Recent trends in wastewater treatments involve utilizing systems which are more energy efficient, less cumbersome and more cost effective. Development of nanofiltration technologies is a major new trend in water and wastewater treatments which are been explored. Nanoadsorbents produced from natural materials such as microbial polysaccharides are currently regarded as one of the effcient and cost-effective workable methods of water and wastewater treatment [51]. Membrane technologies utilizing these types of systems are effective in membrane bioreactors while been used in ultrafiltration processes. This trend is depicted in Fig. 5.6.

A challenging and time-consuming aspect of wastewater treatments is removing heavy metals from the system. Although microbial polysaccharides have been used in the past in the removal of these metals from water through processes like sequestration, their full involvement in bioremediation is attracting more studies in recent years. If fully understood, it will go a long way in treating industrial wastewaters, which are a primary source of ecological worries globally.

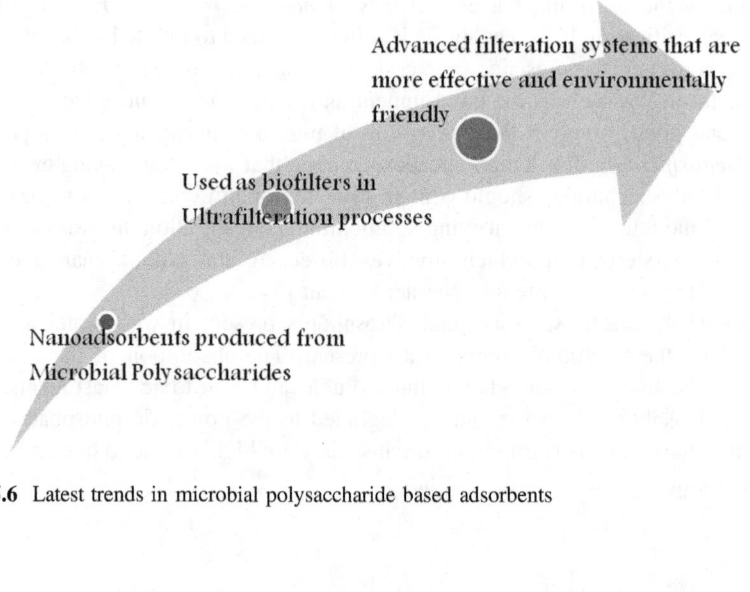

Advanced filteration systems that are more effective and environmentally friendly

Used as biofilters in Ultrafilteration processes

Nanoadsorbents produced from Microbial Polysaccharides

Fig. 5.6 Latest trends in microbial polysaccharide based adsorbents

5.10 Conclusion

The role microbial polysaccharides play in nature cannot be over emphasized. Their availability, as well as ease of production, has made their application tremendous. Although they have been shown to have varied applications in wastewater treatments, their utilization has not been extensively employed. With wastewater treatment processes all over the world being a very expensive and time-consuming process, microbial polysaccharides have shown that they possess great ability in making this cumbersome process less time consuming and more efficient. Microbial polysaccharide as biosorbents have shown remarkable applications in the treatment of textile wastewater which is one of the most difficult wastewaters to treat, due to the presence of both heavy metals and dyes.

The shortage of potable water in developing countries which is partly attributed to low level of wastewater reuse, due to low or unavailable wastewater treatment systems, can be efficiently and economically solved by using systems which incorporate in their operation as treatment regimen in microbial polysaccharides water and wastewater treatment.

Acknowledgements I appreciate Dr. O. Osuolale and my postgraduates students, Jeremiah Ogah and Ifeoluwa Gbala for their contributions. My appreciation also goes to the University of Ilorin, Nigeria, which provided a good working environment to carry out my research works.

References

1. U.U. Nwodo, E. Green, A.I. Okoh, Bacterial exopolysaccharides: functionality and prospects. Int. J. Mol. Sci. **13**, 14002–14015 (2012)
2. M. Indira, T.C. Venkateswarulu, K. Chakravarthy, R.A. Ranganadha, B.D. John, P.K. Vidya, Morphological and biochemical characterization of exopolysaccharide producing bacteria isolated from dairy effluent. J. Pharm. Sci. Res. **8**, 88–91 (2016)
3. A. Poli, P. Donato, G.R. Abbamondi, B. Nicolaus, Synthesis, production, and biotechnological applications of exopolysaccharides and polyhydroxyalkanoates. Archaea (2011). doi:10.1155/2011/693253
4. E.T. Oner, *Microbial production of extracellular polysaccharides from biomass* (Springer, Berlin, 2013)
5. T. Liang, S. Wang, Recent advances in exopolysaccharides from paenibacillus spp.: production, isolation, structure, and bioactivities. Mar. Drugs **13**, 1847–1863 (2015)
6. K.V. Madhuri, K.V. Prabhakar, Recent trends in the characterization of microbial exopolysaccharides. Orient. J. Chem. **30**, 895–904 (2014)
7. R.S. Al-Wasify, A.A. Al-Sayed, S.M. Saleh, A.M. Aboelwafa, Bacterial exopolysaccharides as new natural coagulants for surface water treatment. Int. J. Pharm. Tech. Res. **8**, 198–207 (2015)
8. T.K. Singh, Microbial extracellular polymeric substances: production, isolation and applications. IOSR J. Pharm. **2**, 276–281 (2012)
9. A. Becker, F. Katzen, A. Pühler, L. Ielpi, Xanthan gum biosynthesis and application: a biochemical/genetic perspective. Appl. Microbiol. Biotechnol. **50**, 145–152 (1998)
10. G. Morris, S. Harding, *Polysaccharides, Microbial* (Elsevier, 2009), pp. 482–494

11. A. Patel, J.B. Prajapati, Food and health applications of exopolysaccharides produced by lactic acid bacteria. Adv. Dairy Res. 1 (2013). doi:10.4172/2329-4888X.1000107
12. J.W. Costerton, K.J. Cheng, G.G. Geesey, T.I. Ladd, J.C. Nickel, M. Dasgupta, T.J. Marrie, Bacterial biofilms in nature and disease. Annu. Rev. Microbiol. 41, 435–464 (1987)
13. E. Tok, B. Aslim, Cholesterol removal by some lactic acid bacteria that can be used as probiotic. Microbiol. Immunol. 54, 257–264 (2010)
14. R.M. Jain, K. Mody, A. Mishra, B. Jha, Isolation and structural characterization of biosurfactant produced by an alkaliphilic bacterium Cronobacter sakazakii isolated from oil contaminated wastewater. Carbohydr. Polym. 87, 2320–2326 (2012)
15. C. Calvo, M. Manzanera, G.A. Silva-Castro, I. Uad, J. González-López, Application of bioemulsifiers in soil oil bioremediation processes. Future prospects. Sci. Total Environ. 407, 3634–3640 (2009)
16. J.M. Campos, T.A. Lucia, L.A. Sarubbo, J.M. de Luna, R.D. Rufino, I.M. Banat, Microbial Biosurfactants as Additives for Food Industries. Biotechnol. Prog. 29, 1097–1108 (2013)
17. I.M. Banat, A. Franzetti, I. Gandolfi, G. Bestetti, M.G. Martinotti, L. Fracchia, T.J. Smyth, R. Marchant, Microbial biosurfactants production, applications and future potential. Appl. Microbiol. Biotechnol. 87, 427–444 (2010)
18. R. Kaur, J. Singh, R. Khare, A. Ali, Biosorption the possible alternative to existing conventional technologies for sequestering heavy metal ions from aqueous streams: a review. Univers. J. Environ. Res. Technol. 2, 325–335 (2012)
19. F. Fu, Q. Wang, Removal of heavy metal ions from wastewaters: a review. J. Environ. Management 92, 407–418 (2011)
20. O. Abdi, M. Kazemia, A review study of biosorption of heavy metals and comparison between different biosorbents. J. Mater. Environ. Sci. 6, 1386–1399 (2015)
21. H. Khakpour, H. Younesi, M.M. Hosseini, Two-stage biosorption of selenium from aqueous solution using dried biomass of the baker's yeast Saccharomyces cerevisiae. J. Environ. Chem. Eng. 2, 532–542 (2014)
22. G.M. Gadd, Biosorption: critical review of scientific rationale, environmental importance and significance for pollution treatment. J. Chem. Technol. Biotechnol. 84, 13–28 (2009)
23. S.S. Ahluwalia, D. Goyal, Microbial and plant derived biomass for removal of heavy metals from wastewater. Biores. Technol. 98, 2243–2257 (2007)
24. N. Das, R. Vimala, P. Karthika, Biosorption of heavy metals- An overview. Indian J. Biotechnol. 7, 159–169 (2008)
25. G.Z. Kyzas, J. Fu, K.A. Matis, The change from past to future for adsorbent materials in treatment of dyeing wastewaters. Materials 6, 5131–5158 (2013)
26. G. Crini, Recent developments in polysaccharide-based materials used as adsorbents in wastewater treatment. Prog. Polym. Sci. 30, 38–70 (2005)
27. Z. Ding, I. Bourven, G. Guibaud, E.D. van Hullebusch, A. Panico, F. Pirozzi, G. Esposito, Role of extracellular polymeric substances (EPS) production in bioaggregation: application to wastewater treatment. Appl. Microbiol. Biotechnol. (2015). doi:10.1007/s00253-015-6964-8
28. S. Andersson, Characterization of bacterial biofilms for wastewater treatment, 2009, Printed by Universitets service US-AB, Drottning Kristinas väg 53B SE-100 44 Stockholm, Sweden
29. B. Vu, M. Chen, R.J. Crawford, E.P. Ivanova, Bacterial extracellular polysaccharides involved in biofilm formation. Molecules 14, 2535–2554 (2009). doi:10.3390/molecules14072535
30. M.B.I. Kassim, Production and characterization of the polysaccharide "xanthan gum" by a local isolate of the bacterium Xanthomonas campestris. Afr. J. Biotechnol. 10, 16924–16928 (2011)
31. A.K. Patel, P. Michaud, R.R. Singhania, Polysaccharides from probiotics as food additives. Food Technol. Biotechnol. 48, 451–463 (2010)
32. S. Meisen, J. Wingender, U. Telgheder, Analysis of microbial extracellular polysaccharides in biofilms by HPLC. Part I: development of the analytical method using two complementary stationary phases. Nal. Bioanal. Chem. 391, 993–1002 (2008)

33. S.R. Chowdhury, S. Manna, P. Saha, R.K. Basak, R. Sen, D. Roy, B. Adhikari, Composition analysis and material characterization of an emulsifying extracellular polysaccharide (EPS) produced by Bacillus megaterium RB-05: a hydrodynamic sediment-attached isolate of freshwater origin. J. Appl. Microbiol. **111**, 1381–1393 (2011)

34. A. Nanda, C.M. Raghavan, Production and characterization of exopolysacharides (EPS) from the bacteria isolated from Pharma lab sinks. Int. J. Pharm. Tech. Res. **6**, 1301–1305 (2014)

35. M.B. Lilledahl, B.T. Stokke, Novel imaging technologies for characterization of microbial extracellular polysaccharides. Front. Microbiol. **6**, 1–12 (2015)

36. H. Su, Z. Chen, S. Liu, L. Qiao, X. Chen, H. He, X. Zhao, B. Zhou, Y. Zhang, Characterization of bacterial polysaccharide capsules and detection in the presence of deliquescent water by atomic force microscopy. Appl. Environ. Microbiol. 78 (2012)

37. G. Gonzalez-Gil, L. Thomas, A. Emwas, P.N.L. Lens, P.E. Saikaly, NMR and MALDI-TOF MS based characterization of exopolysaccharides in anaerobic microbial aggregates from full-scale reactors. Sci. Rep. 5 (2015) doi:10.1038/srep14316

38. J. Allen, Microbial polysaccharides: application, production and features (2013), http://www.biologydiscussion.com Accessed 25 Apr 2016

39. F. Donot, A. Fontana, J.C. Baccou, C. Schorr-Galindo, Microbial exopolysaccharides: main examples of synthesis, excretion, genetics and extraction. Carbohyd. Polym. **87**, 951–962 (2012)

40. S. Lee, R. John, F. Mei, C. Chong, A review on application of flocculants in wastewater treatment. Proc. Safety Environ. Protect. **92**, 489–508 (2014)

41. F. Freitas, V. Alves, A.M. Reis, Advances in bacterial exopolysaccharides: from production to biotechnological applications. Trends Biotechnol. **29**, 388–398 (2011)

42. O. Ates, Systems biology of microbial exopolysaccharides production. Front. Bioeng. biotechnol **3**, 1–16 (2015). doi:10.3389/fbioe.2015.00200

43. T. Todhanakasem, Microbial biofilm in the industry. Afr. J. Microbiol. Res. **7**, 1625–1634 (2013)

44. M.A. Kumar, K.T.K. Anandapandian, K. Parthiban, Production and characterization of exopolysaccharides(eps) from biofilm forming marine bacterium. Braz. Arch. Biol. Technol. **54**, 259–265 (2011)

45. I. Sutherland, Biofilm exopolysaccharides: a strong and sticky Framework. Microbiol. **147**, 3–9 (2001)

46. L. Pierre, L. Cécile, D.M. Patrick, Exopolysaccharides of the biofilm matrix: a complex biophysical world, 2012

47. V. Barbara, C. Miao, J. Russell, I. Elena, Xanthan gum biosynthesis and application: a biochemical/genetic perspective. Appl. Microbiol. Biotechnol. **50**, 145–152 (1998)

48. M. Kostakioti, M. Hadjifrangiskou, S. Hultgren, Bacterial biofilms: development, dispersal, and therapeutic strategies in the dawn of the postantibiotic era. Cold Spring Harb. Perspect. Med. 3 (2013) doi:10.1101/cshperspect.a010306

49. M. Ciszek-Lenda, Biological functions of exopolysaccharidesfrom probiotic bacteria. Centr. Eur. J. Immunol. **36**, 51–55 (2011)

50. K. Sugimoto, Biofilm as a new bio-material. Innovatie, in: https://www.rvo.nl/sites/default/files/Biofilm%20Japan.pdf, (Ed.), Attaché Tokio, (2013)

51. Y.G. Maksimova, Microbial biofilms in biotechnological processes. Appl. Biochem. Microbiol. **50**, 750–760 (2013)

52. K. Sambanthamoorthy, F. X., R. Patel, S. Patel, C. Paranavitana, Antimicrobial and antibiofilm potential of biosurfactants isolated from lactobacilli against multi-drug-resistant pathogens. BMC Microbiol. **14** (2014). doi:10.1186/1471-2180-1114-1197

53. R. Donlan, Biofilms: microbial life on surfaces. Emerg. Infect. Diseases **8**, 881–889 (2002)

54. J. Kloc, I. Gonzalez, The study of biological wastewater treatment through biofilm development on synthetic material vs. membranes, (Worcester Polytechnic Institute, Massachusetts, 2012)

55. N. Qureshi, B. Annous, T. Ezeji, Biofilm reactors for industrial bioconversion processes: employing potential of enhanced reaction rates. Microb. Cell Fact. **4**, 24 (2005)

56. G.M. Teitzel, M.R. Parsek, Heavy metal resistance of biofilm and planktonic Pseudomonas aeruginosa. Appl. Environ. Microbiol. **69**, 2313–2320 (2003)
57. S. Vijayakumar, V. Saravanan, Biosurfactants-types, sources and applications. Res. J. Microbiol. **10**, 181–192 (2015)
58. S. Andersson, *Characterization of bacterial biofilms for wastewater treatment* (Universitets service US-AB, Drottning Kristinas väg Stockholm, Sweden, 2009)
59. V. Sharma, A. Sharma, Nanotechnology: an emerging future trend in wastewater treatment with its innovative products and processes. Int J Enhanc. Res. Sci Technol. Eng. **1**, 1–8 (2012)
60. G.C. Delzer, S.W. McKenzie, Five-day biochemical oxygen demand: U.S. geological survey techniques of water-resources investigations, 2003, book 9, chap. A7, section 7.0, November, accessed last date from http://pubs.water.usgs.gov/twri9A/
61. Environmental Protection Agency, *Chemical Contaminant Rules*. (2016) https://www.epa.gov/dwreginfo/chemical-contaminant-rules Accessed 12 Jan 2016
62. T.B.S. Prakasam, R.C. Loehr, Author links open the overlay panel. Numbers correspond to the affiliation list which can be exposed by using the show more link. Water Res. **6**, 859–869 (1972)

Printed in the United States
By Bookmasters